ENERGY SCIENCE, ENGINEERING AND TECHNOLOGY

SHARED SOLAR POWER

PROJECT GUIDES FOR COMMUNITIES

ENERGY SCIENCE, ENGINEERING AND TECHNOLOGY

Additional books in this series can be found on Nova's website under the Series tab.

Additional e-books in this series can be found on Nova's website under the e-book tab.

ENERGY POLICIES, POLITICS AND PRICES

Additional books in this series can be found on Nova's website under the Series tab.

Additional e-books in this series can be found on Nova's website under the e-book tab.

ENERGY SCIENCE, ENGINEERING AND TECHNOLOGY

SHARED SOLAR POWER

PROJECT GUIDES FOR COMMUNITIES

EMILY J. ROSE
AND
DUNCAN MARKS
EDITORS

nova publishers
New York

For permission to use material from this book please contact us:
Telephone 631-231-7269; Fax 631-231-8175
Web Site: http://www.novapublishers.com

Library of Congress Cataloging-in-Publication Data

ISBN: 978-1-62257-527-5

Published by Nova Science Publishers, Inc. † New York

CONTENTS

PREFACE

In communities across the United States, people are seeking alternatives to conventional energy sources. Whether they aim to increase energy independence, hedge against rising fuel costs, cut carbon emissions, or provide jobs, people are looking to community-scale renewable energy projects for solutions. Falling costs and creative new financing models have made solar projects, including community shared solar projects, more financially feasible. "Community shared solar" is defined as a solar-electric system that provides power and/or financial benefit to multiple community members. This book provides current detail on projects designed to increase access to solar energy and to reduce up-front costs for participants.

Chapter 1 – In communities across the United States, people are seeking alternatives to conventional energy sources. Whether they aim to increase energy independence, hedge against rising fuel costs, cut carbon emissions, or provide jobs, people are looking to community-scale renewable energy projects for solutions. Falling costs and creative new financing models have made solar projects—including community shared solar projects—more financially feasible.

This guide is a resource for those who want to develop community shared solar projects, from community organizers or solar energy advocates to government officials or utility managers. By exploring the range of incentives and policies while providing examples of operational community shared solar projects, this guide will help communities plan and implement successful energy projects. In addition, by highlighting some policy best practices, this guide suggests changes in the regulatory landscape that could significantly boost community shared solar installations across the nation.

Chapter 2 – The first "Solarize" campaign started as a grassroots effort to help residents of Portland, Oregon, overcome the financial and logistical barriers to installing solar power. What began in one neighborhood as "Solarize Southeast!" quickly caught on with residents across the city. With support from the U.S. Department of Energy's (DOE) Solar America Communities program, the City of Portland Bureau of Planning and Sustainability partnered with neighborhood coalition offices, Solar Oregon (the state American Solar Energy Society chapter) and Energy Trust of Oregon to provide community organizing, technical assistance, project management, and rebates in a wildly popular grassroots-driven program. After three years of Solarize campaigns, Portland has added over 1.7 MW of distributed photovoltaics (PV) and established a strong, steady solar installation economy.

Since the publication of the first Solarize Guidebook in 2011, dozens of communities, companies and contractors across the U.S. have launched their own versions of a neighborhood collective purchasing program. With installed costs for behind-the-meter (distributed) solar dropping 17% in 2010 and continuing to fall in 2011, the residential PV market in the U.S. is poised to continue expansion and Solarize campaigns can accelerate this growth.

In: Shared Solar Power
Editors: E. J. Rose and D. Marks
ISBN 978-1-62257-527-5

Chapter 1

A GUIDE TO COMMUNITY SHARED SOLAR: UTILITY, PRIVATE, AND NONPROFIT PROJECT DEVELOPMENT*

U. S. Department of Energy

ACKNOWLEDGMENTS

This guide is an updated version of the original Guide to Community Solar, published November 2010 (see www.nrel.gov/docs/fy11osti/49930.pdf), which was developed for the National Renewable Energy Laboratory by Northwest Sustainable Energy for Economic Development, Keyes and Fox, Stoel Rives, and the Bonneville Environmental Foundation. This guide builds on the research and writing from the Northwest Community Solar Guide, published by Bonneville Environmental Foundation and Northwest SEED.

* This is an edited, reformatted and augmented version of the U.S. Department of Energy, NREL, AGG-2-22125-01, dated May 2012.

SECTION 1: INTRODUCTION

Purpose

In communities across the United States, people are seeking alternatives to conventional energy sources. Whether they aim to increase energy independence, hedge against rising fuel costs, cut carbon emissions, or provide jobs, people are looking to community-scale renewable energy projects for solutions. Falling costs and creative new financing models have made solar projects—including community shared solar projects—more financially feasible.

This guide is a resource for those who want to develop community shared solar projects, from community organizers or solar energy advocates to government officials or utility managers. By exploring the range of incentives and policies while providing examples of operational community shared solar projects, this guide will help communities plan and implement successful energy projects. In addition, by highlighting some policy best practices, this guide suggests changes in the regulatory landscape that could significantly boost community shared solar installations across the nation.

How to Use This Guide

The information in this guide is organized around three sponsorship models: utility projects, special purpose entity projects, and nonprofit projects. The guide begins with examples of the three project sponsorship models, discussing the legal and financial implications of each model. This is followed by a discussion of state policies that encourage community shared solar. The guide then reviews some of the tax and financing issues that impact community shared solar projects. While the guide cannot offer legal or tax advice, the authors hope to provide an outline of the legal hurdles that every project organizer should consider. Finally, Section 6, Getting Started provides readers with practical tools and tips for planning their own projects. The Appendices provide a more detailed comparison of business structures suitable for special purpose entities pursuing solar projects and the Interstate Renewable Energy Council's Model Community Renewables Program Rules.

As with the first version of this guide, the case studies have been provided by the program sponsors or developers and have not been independently verified by the authors or by NREL. Please contact the program sponsor for further information.

This guide cannot possibly describe all available incentives or cite all the examples of community shared solar efforts nationwide. For information regarding the most recent developments, see Section 7, Resources.

Why "Community Shared" Solar?

For the purpose of this guide, "community shared solar" is defined as a solar-electric system that provides power and/or financial benefit to multiple community members. Community shared solar advocates recognize that the on-site solar market comprises only one part of the total market for solar energy. A 2008 study by the National Renewable Energy Laboratory (NREL) found that only 22 to 27% of residential rooftop area is suitable for hosting an on-site photovoltaic (PV) system.[1] Community options expand access to solar power for renters, those with shaded roofs, and those who choose not to install a residential system on their home for financial or other reasons. As a group, ratepayers and tax payers fund solar incentive programs. Accordingly, as a matter of equity, solar energy programs should be designed in a manner that allows all contributors to participate.

This guide focuses on projects designed to increase access to solar energy and to reduce up-front costs for participants. Secondary goals met by many community shared solar projects include:

- Improved economies of scale
- Optimal project siting
- Increased public understanding of solar energy
- Local job generation
- Opportunity to test new models of marketing, project financing, and service delivery.

Creative mechanisms to foster greater solar energy project deployment are not limited to those described in this guide. Readers may be interested in investigating the following efforts that employ some elements of community shared solar:

- Volume purchasing efforts, such as those in Portland, OR (Solarize Portland!) and nationwide (One Block Off the Grid)

- Solar services co-ops such as Cooperative Community Energy, CA
- Utility-owned distributed generation on customer rooftops, such as the Arizona Public Service Community Power Project.

Definition of Key Terms

The following terms are defined in the context of community shared solar.

Renewable Energy Certificates (RECs, carbon offsets, or green tags)

A renewable energy facility produces two distinct products. The first is electricity. The second is the package of environmental benefits resulting from not generating the same electricity—and emissions—from a conventional gas or coal-fired power plant. These environmental benefits can be packaged into a REC and sold separately from the electrical power. A REC represents the collective environmental benefits, such as avoided mercury, carbon dioxide (CO_2), and other environmentally harmful pollutants, as a result of generating one megawatt-hour (MWh) of renewable energy.

In most cases, RECs are sold on a per MWh basis. However, some project organizers choose to sell all future rights to RECs up front, on a per-installed-watt basis, effectively capturing an installation rebate and forgoing any future revenue from REC sales.

Net metering

Most on-site renewable energy systems use net metering to account for the value of the electricity produced when production is greater than demand. Net metering allows customers to bank this excess electric generation on the grid, usually in the form of kilowatt-hour (kWh) credits during a given period. Whenever the customer's system is producing more energy than the customer is consuming, the excess energy flows to the grid and the customer's meter "runs backwards." This results in the customer purchasing fewer kilowatt-hours from the utility, so the electricity produced from the renewable energy system can be valued at the retail price of power. Most utilities have a size limit for net metering. Community shared solar project organizers should be sure to check before assuming participants in a community shared solar system can net meter. It may be that some alternative arrangement, such as group billing or joint ownership, is used to account for the value of the electricity produced by a community shared solar project.

Tax appetite

Individuals and businesses can reduce the amount of taxes owed by using tax credits. For a tax credit to have any value, though, the individual or business must actually owe taxes. If the individual or business is tax exempt or does not have sufficient income to need tax relief, the tax credits have no value. Individuals or businesses that can use tax credits to reduce the amount they owe in taxes are said to have a "tax appetite." For example, public and nonprofit organizations are tax exempt, and therefore, do not have a tax appetite. In addition, taxpaying entities might be eligible to use tax-based incentives, but have insufficient tax appetite to make full use of them.

Investment Tax Credit (ITC)

Section 48 of the Internal Revenue Code defines the federal ITC. The ITC allows commercial, industrial, and utility owners of PV systems to take a one-time tax credit equivalent to 30% of qualified installed costs. There is also a federal residential renewable energy tax credit (Internal Revenue Code Section 25D), but the residential tax credit requires that the PV system be installed on a home the taxpayer owns and uses as a residence, thus it would rarely, if ever, be applicable to community shared solar projects.

Power purchase agreement (PPA)

A PPA is an agreement between a wholesale energy producer and a utility under which the utility agrees to purchase power. The PPA includes details such as the rates paid for electricity and the time period during which it will be purchased. Sometimes, the term PPA or "third-party PPA" is used to describe the agreement between the system owner and the on-site system host, under which the host purchases power from the system. This arrangement is not explicitly allowed in all states; in some states, it may subject the system owner to regulation as a utility. To avoid confusion, in this guide, a PPA refers only to an agreement by a utility to purchase power from the solar system owner.

Solar services agreement (SSA)

A solar services agreement is an agreement between the system owner and the system site host, for the provision of solar power and associated services. The system owner designs, installs, and maintains the system (a set of solar services) and signs an agreement with the host to continue to provide maintenance and solar power. The agreement is sometimes referred to as a PPA, but in this guide, we use the term SSA to indicate that the agreement between the system owner and the system site host is more than a power purchase: it is an agreement that the system owner will provide specific services to ensure continued solar power.

Securities

A security is an investment instrument issued by a corporation, government, or other organization that offers evidence of debt or equity. Any transaction that involves an investment of money in an enterprise, with an expectation of profits to be earned through the efforts of someone other than the investor, is a transaction involving a security. Community shared solar organizers must be sure to comply with both state and federal securities regulations, and avoid inadvertently offering a security. For more information on securities, see Section 4, Tax Policies and Incentives.

SECTION 2: COMMUNITY SHARED SOLAR PROJECT MODELS

People have many reasons for organizing or participating in a community shared solar project. Just as their motives vary, so do the possible project models, each with a unique set of costs, benefits, responsibilities, and rewards. This section reviews several project models:

- **Utility-Sponsored Model:** A utility owns or operates a project that is open to voluntary ratepayer participation.
- **Special Purpose Entity (SPE) Model:** Individuals join in a business enterprise to develop a community shared solar project.
- **Nonprofit Model:** A charitable nonprofit corporation administers a community shared solar project on behalf of donors or members.

The authors of this guide illustrate pros and cons of different sponsorship models, as well as variations within project models, so that project planners can select the model and variations that best suit their situation and goals. Before selecting a project model, every planner should consider the issues below.

- **Allocation of Costs and Benefits:** Who will pay to plan, construct, and operate the solar system? Who will have rights to benefits, including the electricity produced, RECs, revenue from electricity sales, tax benefits, other incentives, and ownership of the project's assets (such as the solar system itself)?

- **Financial and Tax Considerations:** Will money be raised through a solar fee on electricity bills, by equity or debt financing of a business entity, through charitable donations, or other options? What kind of tax implications will there be for participants—e.g., will the project generate taxable income for participants? Will it generate tax credits or deductions for participants?
- **Other Legal Issues:** How will the project design address securities regulation, utilities regulation, business regulation, and the complexity of agreements between various project participants?

The chart on the following page compares aspects of the three sponsorship models.

Comparison of Models

	Utility	**Special Purpose Entity**	**Nonprofit**
Owned By	Utility or third party	SPE members	Nonprofit
Financed By	Utility, grants, ratepayer subscriptions	Member investments, grants, incentives	Memberships, donor contributions, grants
Hosted By	Utility or third party	Third party	Nonprofit
Subscriber Profile	Electric rate payers of the utility	Community investors	Donors, members
Subscriber Motive	Offset personal electricity use	Return on investment; offset personal electricity use	Return on investment; philanthropy
Long-term Strategy of Sponsor	Offer solar options; add solar generation (possibly for Renewa-ble Portfolio Standard)	Sell system to host; retain for electricity production	Retain for electricity production for life of system
Examples	• Sacramento Municipal Utility District – SolarShares Program • Tucson Electric Power – Bright Tucson Program	• University Park Community Solar, LLC • Clean Energy Collective, LLC • Island Community Solar, LLC	• Winthrop Community Solar Project • Solar for Sakai

Utility-Sponsored Model

For communities desiring to organize a community shared solar project, the local electric utility is a good place to start. First of all, utilities are likely to have the legal, financial, and program management infrastructure to handle organizing and implementing a community shared solar project. Second, many utilities are actually governed by the member customers and can be directed to pursue projects on members' behalf. Fully one-fourth of Americans own their own electric power company through co-ops, or city- or county-owned utilities.[2] In general, publicly owned utilities have taken the lead in deploying community shared solar projects. Even when the utility is investor-owned or privately held, it may wish to expand customer choice with an option for community shared solar power.[3]

Overview

In most utility-sponsored projects, utility customers participate by contributing either an up-front or ongoing payment to support a solar project. In exchange, customers receive a payment or credit on their electric bills that is proportional to 1) their contribution and 2) how much electricity the solar project produces. Usually, the utility or some identified third party owns the solar system itself. The participating customer has no ownership stake in the solar system. Rather, the customer buys rights to the benefits of the energy produced by the system. Note that utility-sponsored community shared solar programs differ from traditional utility "green power" programs in that "green power" programs sell RECs from various renewable energy resources and generally do not act as a hedge against rising electric costs; utility community shared solar programs sell energy or rights to energy from specific solar installations, with or without the RECs, at a rate that is generally locked in for a period of many years.

Utility-sponsored programs can help make solar power more accessible by decreasing the amount of the purchase required, and by enabling customers to purchase solar electricity in monthly increments. Both Sacramento Municipal Utility District's SolarShares and Tucson Electric Power's Bright Tucson programs allow customers to participate in community shared solar on a monthly basis.

Since 2008, investor-owned utilities have been eligible to use the commercial ITC on qualifying public utility property. And as taxpaying entities, the utilities may have the tax appetite to make use of them. However, normalization accounting rules limit regulated utilities' flexibility in maximizing the value of these tax benefits compared to other private developers. Normalization rules require regulated utilities to spread the benefits of investment tax credits throughout the useful life of the solar project in the rate-making process. The utility's incentive for investment is the difference between the value it receives from the tax credit up front and the value it passes on to customers over time (i.e., the time value of money). Private developers have the flexibility to pass on the benefits of the ITC sooner, which can give them a price advantage over utility solar projects.[4]

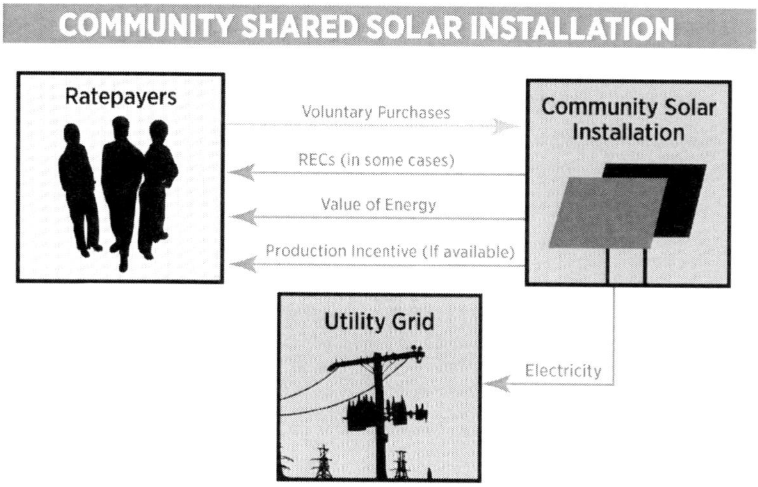

Other legal issues for utility-sponsored projects include the following:

- **Securities Compliance.** In designing mechanisms for customer participation in solar projects, utilities must be careful to comply with state and federal securities regulations. This requires carefully considering what benefit a customer-participant receives in exchange for a financial contribution to the project and how the project is marketed. For example, customer participants may be offered ownership stakes in the solar system itself or just the rights to certain benefits from the energy produced (such as credit on their electric bills, RECs, or access to a special electric rate). However, regardless of how the program is marketed, depending on your state, the receipt of credits on electric bills or other benefits may constitute a return on an investment and fall within the blue sky laws (state laws that regulate the offering and sale of securities).

- **Allocation of Incentives.** In addition to federal tax incentives, a utility-sponsored project might be eligible for various state incentive programs that provide cash benefits or savings to the project. The utility must consider whether and how these incentives will be passed on to customer participants and the tax implications of how the incentives are handled. For example, in Washington State, participants in a utility-sponsored program are eligible for production incentives. While the state Department of Revenue has ruled that the incentive is not taxable, the IRS has not ruled definitively on whether subsidies for solar PV in community shared solar installations are taxable income, although the precedent is that subsidies for energy conservation measures are not taxable.[5]
- **RECs.** Customer participants in utility-sponsored projects often desire to claim the environmental benefits of using solar energy. Participants can only make such a claim if they receive RECs or the utility retires the RECs on the participants' behalf. If the utility keeps the RECs for any reason, including Renewable Portfolio Standard compliance, only the utility can make environmental claims related to the solar system. The utility-sponsored project should consider and make explicit how RECs are allocated.

From a participant perspective, the tax implications are minimal. Bill credits for the value of electricity are not generally taxed; at the same time, participants in a utility-sponsored project are not eligible for the federal investment tax credit. The relative ease of participating in a utility-sponsored project may offset some of the foregone tax incentives available under other community shared solar ownership models.

Examples of Utility-Sponsored Projects

The following examples highlight some of the project options available to those planning a utility-sponsored project.

Sacramento Municipal Utility District (SMUD): SolarShares Program

SMUD's SolarShares Program allows customers who cannot or choose not to acquire PV systems of their own to purchase solar power directly from SMUD while achieving net metering benefits comparable to behind-the-meter PV. SMUD buys the output of local, community-scale photovoltaic systems under 20-year PPAs and then resells the solar power to participating customers. Bill credits equivalent to the amount of energy the customer buys from the SolarShares system are credited to the customer through virtual net metering and are equivalent in value to the bill credits received by a customer with behind-the-meter PV—i.e., full retail price per kWh. The program is subsidized with SB1[6] surcharge funds, which allows SMUD to sell the power for less than the PPA purchase price. SMUD retains the renewable energy credits and is able to count up to 25 MW of SolarShares projects toward its 125-MW SB1 goal. SolarShares' business goals are to make solar benefits available to all SMUD ratepayers, to contribute to achieving SMUD's 125-MW SB1 goal, and to gather pricing and marketing experience that could lead to a sustainable solar enterprise for SMUD beyond the current, mandated incentive program.

SolarShares began in mid-2008 with a 1-MW system constructed by enXco at a leased site in Wilton. The system has thus far produced an average 1,745 MWh per year, of which about 86% has been sold to SolarShares participants. Intensified marketing in Q4 2011 succeeded in moving the percentage sold toward the program's 95% goal. The program has maintained stable enrollment of around 600 customers throughout its three-year life, with most dropouts attributable to customers moving out of the District. Market research conducted in mid-2009 confirmed that most SolarShares customers are satisfied with the program (75% positive responses) and would recommend it to others (85% positive responses).

Customers pay a fixed monthly fee, based on both their average electricity consumption and the amount of PV to which they want to subscribe (from 0.5 to 4 kW). SMUD is exploring the marketing advantages of changing this pricing structure to a flat fixed fee per kWh, allowing customers to purchase in packets of 1,000 kWh/year. Once enrolled, customers are locked in at the fixed monthly fee, for as long as they wish to participate. They receive monthly kWh credits for the estimated output of their solar subscription. Although customers currently pay a premium for solar energy, the effective rate for solar is locked in when they enroll, which maintains the ability of solar to act as a hedge against future price increases. SMUD is making plans for expansion of up to 25 MW by the end of 2016. An RFP for a second megawatt was released in Q3 2011, and the next 1-MW project is scheduled for completion in Q3 2012. The PPA price for the second MW will be blended with the price for the original system to yield a lower participation fee for both existing and new program subscribers. Depending on market response to the second project, SMUD will probably seek to expand the program by larger increments in the future (the enabling legislation caps projects at 5 MW each).

- **Program Highlights**
  - *System Owner:* enXco, with SMUD purchasing 100% of the output under a 20-year PPA
  - *Installed Capacity:* 1 MW
  - *Participant Agreement:* Customers pay a fixed monthly fee in return for a kWh credit. Credit varies monthly, as solar output varies, so a 12-month consecutive commitment is requested.
  - *Electricity:* The estimated kWh generated by a customer's share is netted against the customer's consumption at home, at the full retail rate.
  - *RECs:* Retained by SMUD
  - *Number of Participants:* Approximately 600
- **Financial Details**
  - *Installed Cost*: NA
  - *Capital Financing*: Handled by third party, enXco
  - *Tax Credits*: 30% federal business investment tax credit taken by enXco, depreciation taken by enXco
  - *Estimated Annual Cost*: Varies by customer size and array size. Output from a 0.5-kW share for the small user will cost $129/year at 2012 prices. As the price for non-solar energy rises, a participant could eventually realize monthly savings on their solar purchase.

For more information: Stephen Frantz, sfrantz@smud.org, (916) 732-5107, www.smud.org/

Photo from Marc Romito, Tucson Electric Power.

Tucson Electric Power: Bright Tucson Community Solar Program

In 2011, Tucson Electric Power launched its Bright Tucson Community Solar Program to create opportunities for customers unable to install traditional distributed solar power. Through the program, customers have the opportunity to purchase solar power in “blocks” of 150 kWh per month. Program participants can choose to purchase some or all of their energy through the program. Each purchased block replaces the charges for an equivalent amount of conventional power. At current rates, the solar block is more expensive by about two cents per kWh, but program blocks are exempt from two surcharges applied to other electric usage. Both these surcharges are adjusted annually to reflect changing energy costs, so the benefit of avoiding them could increase over time. The solar block rate is locked in for 20 years under rules approved by the Arizona Corporation Commission (ACC), offering TEP customers a way to hedge against future rate increases. While blocks purchased through the program will still be subject to non-fuel rate changes, the blocks will not be affected by changes to the base energy rate or renewable energy surcharges.

Tucson Electric Power offers an online solar calculator to help potential participants determine how many blocks to purchase to offset the desired quantity of household electricity use. If the solar energy purchased through the program exceeds actual usage during a monthly billing period, the excess is carried forward to the next billing period as a credit. Any credit remaining after the September billing period will be paid in full as a credit on the next bill.

The first source of solar power for the Bright Tucson Community Solar Program is a 1.6-MW single-axis tracking PV array located in The Solar Zone at the University of Arizona Science and Technology Park. TEP is expanding the program as demand requires through utility-owned systems and power purchase agreements. Currently, program participants have purchased 2.1 MW of community shared solar.

The following details pertain specifically to the first Bright Tucson Community Solar Program solar source, a 1.6-MW single-axis tracking PV array, unless otherwise noted.

- **Program Highlights**
  - *System Owner:* Tucson Electric Power
  - *System Host:* University of Arizona Science and Technology Park
  - *Installed Capacity:* 1.6-MW single-axis tracking PV array
  - *Participant Agreement*: Customers pay a fixed monthly fee per solar block in return for a 150-kWh credit. Any credit remaining after the September billing period will be paid in full as a credit on the next bill.

  - *Electricity:* Each 150-kWh block replaces the charges for an equivalent amount of conventional power at a rate that currently adds $3 per month to the customer's electric bill.
  - *RECs:* Retained by TEP
  - *Number of Participants*: 564 (six are commercial; includes all program solar sources)
- **Financial Details**
  - *Installed Cost*: $4/watt
  - *Capital Financing*: Utility financed
  - *Tax Credits*: For 1.6-MW single-axis tracking array, TEP used levelized ITC. For 2-MW dual-axis tracking array, owner took the Treasury Grant (in lieu of ITC).
  - *Estimated Annual Cost*: $36/year for a monthly 150-kWh block. As the price for non-solar energy rises, participants could eventually realize monthly savings on their solar purchase.

For more information: Marc Romito, mromito@tep.com, www.tep.com/Renewable/Home/Bright

Other Community Shared Solar Projects

United Power, CO; City of Ellensburg, WA; Florida Keys Electric Co-op, FL; Seattle City Light, WA; St. George, UT; City of Ashland, OR; *Coming Soon: San Diego Gas & Electric, CA*

Special Purpose Entity (SPE) Models

To take advantage of the tax incentives available to commercial solar projects, organizers may choose to structure a project as a business. In most states, there is a range of business entities that could be suitable for a participant-owned community shared solar project. (Please see Appendix A for more in-depth descriptions of these business entities.) The main challenges in adapting these commercial solar structures for community shared projects include:

- Fully using available tax benefits when community investors have a limited tax appetite, including a lack of passive income
- Maintaining the community project identity when engaging non-community-based tax-motivated investors

- Working within limits on the number of unaccredited investors if the project is to be exempt under securities laws.

Overview

When a group chooses to develop a community shared solar project as a special purpose entity, it assumes the significant complexity of forming and running a business. The group must navigate the legal and financial hurdles of setting up a business and raising capital, and comply with securities regulation. In addition, it must negotiate contracts among the participant/owners, the site host and the utility; set up legal and financial processes for sharing benefits; and manage business operations.

Given the complexity of forming a business, it is not surprising that many special purpose entities pursuing community shared solar are organized by other existing business entities with legal and financial savvy. Solar installation companies such as My Generation Energy in Massachusetts have successfully created LLCs to purchase solar installations funded by groups of investors. Although this expands the market for solar, the benefits are limited to a small group of tax-motivated investors. In an alternative model, the Clean Energy Collective in Colorado has created a business structure under which participation is offered to an unlimited number of utility customers.

Tax and Finance Issues for Special Purpose Entity Projects

Federal income tax benefits offer significant value for solar projects, but can be challenging for community shared projects to use effectively. Making use of tax credits or losses (from depreciation) requires significant taxable income. Moreover, passive investors in a community shared solar project (investors who do not take an active role in the company or its management) can only apply the ITC to passive income tax liability. As discussed below, most investors in a community shared solar project will likely be passive investors, and few will have passive income. As a result, most individuals cannot fully use federal tax benefits. In this section, we describe the major limitations on using federal tax benefits and outline potential financing structures that accommodate those limitations. However, the descriptions here do not account for the many nuances that might apply to individual projects.

Passive Activity Rules

IRS "passive activity" rules are a major challenge for community-based renewable energy investors trying to use federal tax benefits. In most cases, an individual's investment in a community shared solar project will be considered a passive investment. Passive activity rules allow tax credits or losses generated from passive investment to be used to offset only passive income.[7]

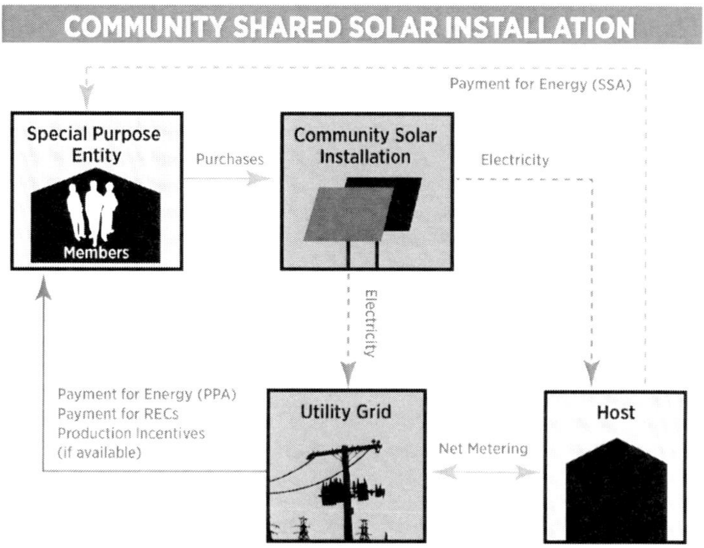

Most individuals primarily have non-passive income, which includes salaries, wages, commissions, self-employment income, taxable social security, and other retirement benefits. Non-passive income also includes portfolio income, such as interest, dividends, annuities, or royalties not derived in the ordinary course of a business. While portfolio income may seem passive, the IRS specifically excludes it from the category of passive income.

Passive income can only be generated by a passive activity. There are only two sources for passive income: a rental activity or a business in which the taxpayer does not materially participate.

"Participation" generally refers to work done in connection with an activity in which the taxpayer owns an interest. To "materially" participate in the trade or business activity (in this case, operation of a solar project) an individual must participate on a regular, continuous, and substantial basis in the operations of the activity. This is a high standard that participants likely will not be able to meet. That means most participants will be passive investors, limited to applying federal tax benefits to passive income. The community shared solar project itself likely will not generate sufficient income to make full use of the ITC or depreciation benefits, at least not in the early years of a project. Therefore, a project intending to rely on federal tax benefits will have to seek participation of an investor with a larger tax appetite.

At-Risk Limitations

In addition to passive activity rules, at-risk rules limit the amount of losses one can claim from most activities. Specifically, one can only claim losses equivalent to one's amount of risk in the activity. The "at-risk" amount generally is the amount of cash and property one contributes to the activity. In addition, any amount borrowed for use in the activity is at-risk, as long as the borrower is personally liable for repayment of the loan or the loan is secured with property not used for the activity. Money contributed from a non-recourse loan is not considered "at-risk."

Securities Regulation

Securities regulations are a major factor in financing structures for the SPE model. To reduce the burden of securities compliance, many small projects seek a private placement exemption to registration requirements. Qualifying for such an exemption requires limiting who can invest in the project (based on assets or income for individuals) and how such an offering can be conducted. The practical effect is to limit the number of middle-income individuals who can invest in a community shared solar project. If a project is designed to produce electricity proportional to the amount used by the participants, securities issues will effectively limit the size of a project. For example, private placement exemption limits the number of "unaccredited" investors to 35 or fewer.[8] A 1-MW solar facility, in contrast, could serve far more participants, perhaps 300 to 500. Therefore, project developers must carefully consider how to reconcile their financing mechanism with the size of their project, the number of participants, and type of participants.

Potential Financing Structures

Special purpose entities need to plan their financing structures carefully. Structures that effectively use the ITC can be complex and tend to mimic the structures used by larger commercial solar projects. For a community SPE, potential financing structures that maximize federal tax incentives include:

- **Self-financing:** This is the simplest option for a community SPE is to finance the project with equity invested by community members. However, in order to fully use federal tax benefits, the SPE needs to have enough community investors that have sufficient tax appetite to use federal tax incentives. Given the passive loss rules and the at-risk limitations discussed above, this is not a realistic goal for community groups consisting of individuals who lack other sources of passive income. That means the project organizers will likely have to make the project economically viable without full use of federal tax incentives (difficult without aid from a state or local incentive of similar value), or will have to use one of the more complex structures such as a flip or a sale/leaseback (described below). This need not take away from the community ownership, if the project can find even one community member with the financial resources and tax appetite to participate as the primary tax investor.

- **Flip Structure:** In this scenario, the community SPE partners with a tax-motivated investor in a new special purpose entity that owns and operates the project. Initially, most of the equity comes from the tax investor and most of the benefit (as much as 99%) would flow to the tax investor. When the tax investor has fully monetized the tax benefits and achieved an agreed-upon rate of return, the allocation of benefits and majority ownership (95%) would "flip" to the community SPE (but not within the first five years). After the flip, the community SPE has the option to buy out all or most of the tax investor's interest in the project at the fair market value of the tax investor's remaining interest. Note that the numbers provided here reflect IRS guidelines on flip structures issued for wind projects claiming the federal production tax credit. Similar rules potentially could apply to solar projects claiming the ITC.
- **Sale/Leaseback:** In this scenario, the community SPE (as the developer of the project, the site host, or both) installs the PV system, sells it to a tax investor and then leases it back. As the lessee, the community SPE is responsible for operating and maintaining the solar system and has the right to sell or use the power. In exchange for use of the solar system, the community lessee makes lease payments to the tax investor (the lessor). The tax investor has rights to federal tax benefits generated by the project and the lease payments. The community SPE may have the option to buy back the project at 100% fair market value after the tax benefits are exhausted.

There are numerous complex legal, financial, and tax issues associated with all of these financing structures. These descriptions do not cover these issues completely. For more information on financing structures, see Section 7, Resources.

Examples of Special Purpose Entity Projects

The following examples represent two possible approaches: a volunteer-led LLC and a business enterprise that partners with utilities to deliver solar to customers. These special purpose entities are structured as LLCs. Although there has been much interest in the possibility of structuring a community shared solar enterprise as a cooperative (co-op), in fact, co-ops are not exempt from the complex securities issues and project organizers have tended to choose to do business as LLCs.[9] Several rural electric co-ops that deliver electricity to customer/members have started community shared solar programs, but the programs are peripheral to the function as consumer co-ops for the distribution of electricity. As in the previous edition of this guide, the descriptions of the programs in the following pages have been provided by the program sponsors or developers and have not been independently verified by the authors or by DOE.

University Park Community Solar LLC, Maryland

The volunteer founders of University Park Community Solar spent more than two years crafting the legal and financial aspects of their business model. With expert consultation, including help from a state senator to change the Maryland net metering law, the volunteers formed a member-managed LLC that will return their investment in five to six years. Within the group, there are both active and passive investors.

Photo from David Brosch, University Park Community Solar, LLC.

A 22-kW system was installed on the roof of a local church in May 2010. The LLC will pass benefits to its members based on revenue from several sources: electricity sold to the church and grid, the auction of RECs, federal tax incentives, and depreciation. The LLC and the Church signed a 20-year agreement detailing the provision of electricity, access to the solar array, maintenance, insurance, and other issues. The host has an option to purchase the system before the 20-year term is up.

To assist in establishing the LLC, the group received pro bono help from the Maryland Intellectual Property Legal Resource Center and paid approximately $12,000 for other legal and accounting expertise. The founders note that initial accounting and legal fees could overwhelm any return to members. Going forward, they plan to handle the accounting and tax paperwork in house as much as possible.

The LLC organizers were careful to obtain legal advice on how to gain an exemption from state and federal SEC filing requirements. The organizers are not all "accredited" investors. In addition, the organizers were required to create lengthy disclosure documents to ensure that investors were fully informed of the risks. Their attorneys advised them to pursue an exemption that restricted them in several aspects, including having fewer than 35 unaccredited investors, keeping the offering private, and limiting membership within the state of Maryland. See Section 5, Securities Compliance, for information about securities compliance and private placement exemptions.

Project founders are looking to expand the model beyond the first site. Additional host sites in Maryland and other states are being explored, including schools, nonprofits, and places of worship. Furthermore, the LLC has offered to share legal and accounting documents with groups around the nation to facilitate the model's replication. The first successful replication was completed in December 2011 by Greenbelt Community Solar, LLC in Greenbelt, Maryland.

- **Program Highlights**
  - *System Owner:* University Park Community Solar, LLC
  - *System Host:* Church of the Brethren, University Park, MD
  - *Installed Capacity:* 22 kW
  - *Participant Agreement:* LLC passes net revenues (after expenses) and tax credits to members
  - *Electricity:* LLC sells power to the church below retail rate; rate escalates approximately 3.5%/year; church net meters and annual net excess generation is compensated by the utility

    - *RECs:* LLC is working to auction RECs independently
    - *Number of Participants*: 35 LLC Members
- **Financial Details**
    - *Installed Cost:* $5.90/watt
    - *Capital Financing:* Project financed with member investments
    - *Tax Credits:* $39,000 ITC (taken as the 1603 Treasury Grant in lieu of a tax credit)
    - *Grants:* $10,000 from state of MD
    - *MACRS:* Will depreciate 85% of cost over six years
    - *Annual Income from Power Sales*: $3,300 in the first year, rising 3.5%/year
    - *Estimated Annual Income from REC Sales:* $7,000 (28 RECs at $250 per MWh)

For more information: David Brosch, davidcbrosch@comcast.net, (301) 779-3168, www.universityparksolar.com

Clean Energy Collective, LLC, Colorado

The Clean Energy Collective (CEC) provides a member-owned model that enables individuals to directly own panels in a community shared solar farm. The CEC works closely with local utilities to create community-scale solar projects that combine the on-bill credits of a utility-owned project with the equivalent tax benefits and rebates of an individually owned solar project. While the 30% investment tax credit is not directly available to individuals who participate in the project, the cost to participate is adjusted to reflect the value of the tax credits. For projects initiated in 2011 or earlier, the CEC took the 1603 Treasury Grant, instead of the ITC, as the initial owner of the array. Portions of the array were then sold to customers at discounted costs (reducing the cost by the proportioned Treasury Grant amount). Customers could not take a tax credit on their purchase because the grant had been taken by the CEC. Both parties are subject to recapture over the first five years if the resulting system is then sold to a disqualified or non-taxpaying entity. Creating this proprietary project model, with ownership, tax and legal considerations, proved challenging.

When individuals purchase panels in the solar farm, the utility credits them for the electricity produced at or above the retail rate using the CEC's RemoteMeter™ software system. The purchase price is as low as $535, depending on location, available rebates, and RECs. For example, in the first project, CEC sold the rights to all future RECs up front on a per-watt basis, offsetting a portion of the installed cost. The benefits of ownership are transferable. If an owner moves within the service territory, the bill credits follow them; if an owner moves out of the territory, the owner can resell ownership to another utility customer or back to the CEC at fair market value, or donate the property to a nonprofit.

The owners must be customers of the electric utility in which the community array is located and their purchase is limited to the number of panels they need to offset 120% of their yearly electric use. These rules ensure that benefits directly accrue to the local utility customers rather than outside investors. The CEC is the management company representing the community owners and maintaining the solar arrays. In order to provide "utility-grade" long-term power to the utility, a percentage of the monthly power credit value and the initial sale price goes toward funding insurance, operations, and maintenance escrows.

The first CEC project is a 78-kW array in the Holy Cross Energy service territory. The CEC leased the land, sold the project to customers, and negotiated a PPA with Holy Cross Energy. The PPA rate paid by Holy Cross will escalate as regular utility rates increase. CEC's RemoteMeter™ system automatically calculates monthly bill credits for customer accounts and integrates directly with the utility's billing system to apply the credits.

In 2011, the CEC completed three more projects, bringing its installed project portfolio to 2.5 MW.

- **Project Highlights – First Project: Mid Valley Metro Solar Array**
  - *System Owner:* Individuals and businesses in Holy Cross Energy utility territory
  - *System Host:* CEC leases site from the Mid Valley Metropolitan District
  - *Installed Capacity:* 78 kW
  - *Participant Agreement:* Minimum $725 purchase (a single panel after rebates and incentives). Panel owners receive monthly credits for the value of the electricity produced for 50 years.
  - *Electricity:* CEC, as agent for its customers, has a PPA with Holy Cross Energy to purchase the power produced. Customers receive the resulting monetary credit on their monthly electric bill.
  - *RECs:* Holy Cross Energy purchased rights to RECs for $500/kW installed (paid up front).
  - *Number of Participants:* 18 customers

- **Financial Details – First Project**
  - *Installed Cost*: $466,000 or $6/watt (Cost to customers: $3.15/watt, includes all rebates, RECs and credits taken by the CEC)
  - *Capital Financing:* Project built with internal CEC private capital, which is paid back as individuals buy in to the project
  - *Federal Tax Credit:* CEC takes the 1603 Treasury Grant and passes the savings to the customer
  - *Rebates:* $1/watt plus $0.50/watt for rights to the RECs from Holy Cross Energy
  - *Estimated Annual Income from Power Sales:* $15,444 ($198/kW), rising as regular rates rise
  - *Simple Payback:* 13.1 years
- **Project Highlights – Subsequent Three Projects**
  - *System Owner:* Individuals, businesses, and educational institutions in various Colorado utility territories
  - *System Host:* CEC leases sites from government and private entities
  - *Installed Capacity:* 858 kW, 1.1 MW, and 498 kW
  - *Participant Agreement:* Minimum purchase ranges from $535 to $756 (a single panel after rebates and incentives). Panel owners receive monthly credits for the value of the electricity produced.
  - *Electricity:* CEC, as agent for its customers, has a PPA with the utility to purchase the power produced, or has an established rate tariff. Customers receive the resulting monetary credit on their monthly electric bill.
  - *RECs:* Utilities purchased rights to RECs for $500/kW installed (paid up front).
  - *Number of Participants*: 400, 500, and 200
- **Financing Details – Subsequent Three Projects**
  - *Installed Cost:* $6/watt, $6/watt, $5.30/watt (cost to customers as low as $3/W includes all rebates, RECs and credits taken by the CEC)
  - *Capital Financing:* Projects built with bridge loan financing from JP Morgan Chase and internal CEC private capital
  - *Federal Tax Credit:* CEC takes the 1603 Treasury Grant and passes the savings to the customer

- *Rebates:* \$1.25/watt to \$1.58/watt, including up-front sale of RECs
- *Estimated Annual Income from Power Sales:* \$172,000, \$220,000 and \$78,300. Rising as regular rates rise
- *Simple Payback:* 12.5 to 15.5 years

For more information: Lauren Suhrbier, *Lauren@easycleanenergy.com* , (970) 319-3939, *www.easycleanenergy.com*

Island Community Solar, LLC, Washington

Inspired by the passage of Washington State's generous production incentive for community shared solar projects (WAC 458-20-273), a group of solar enthusiasts developed a project in their community on Whidbey Island, Washington. Working closely with the local Port District and the utility, Puget Sound Energy, they developed a one-acre "P-Patch" for solar farmers on Port property at Greenbank Farm. The P-Patch consists of six separately metered plots, each capable of hosting approximately 25 kW of ground mounted solar panels. The solar farmers pay rent to the Port and sell power directly to the grid. When the acre is fully built out, it will generate almost enough to match the on-site annual consumption.

In order to capture the investment tax credit, the Whidbey Island group chose to form an LLC, Island Community Solar (ICS). ICS obtained exemption from securities filing requirements under the Federal Intrastate Offering Exemption (Rule 147) and a Washington Small Offering Exemption (WAC 460-44A-504), which prohibits advertising and limits the number of unaccredited investors. After preparing extensive disclosure documents, ICS raised \$430,000 from 36 local members. ICS built 50 kW in two phases, completing the installation in January 2012.

Photo from Linda Irvine, Island Community Solar LLC

ICS projects a positive return on investment over the ten year lease period. The 1603 Treasury Grant enabled the LLC to monetize the investment tax credit. Although most members do not have sufficient tax appetite to use the passive losses from depreciation, they will earn a return from the state production incentive and power sales to the utility.

It may be difficult to replicate or expand this project without policy changes. The expiration of the 1603 Treasury Grant makes it unlikely that the members will be able to monetize future tax credits, because most lack the tax appetite. The sunset of the Washington State production incentive in June 2020 means that every subsequent project has a shorter window of opportunity to earn incentives. Finally, the avoided cost of the power generated is dropping. The utility's PPA rates for 2012 are lower than in 2011, due to many factors including downward pressure on electric prices from an abundance of natural gas, and the discarding of an assumed future cost for carbon.

- **Project Highlights**
  - *System Owner:* Island Community Solar, LLC
  - *System Host:* Port of Coupeville's Greenbank Farm
  - *Installed Capacity:* 50 kW; estimated Production: 52,930 kWh/year
  - *Participant Agreement*: Members receive distributions, profits, and losses in proportion to capital contributions; passive loss limitations apply.
  - *Electricity:* Sold to the utility through a 10 year PPA, escalating 2.5% annually
  - *RECs:* Retained by the owner; no market for solar RECs in WA
  - *Number of Participants:* 36
- **Financial Details**
  - *Installed Cost*: $410,000 installation; $8,000 legal; $5,400/year insurance
  - *Capital Financing*: 100% owner equity
  - *Federal Tax Credit:* $123,000 1603 Treasury Grant
  - *Incentives:* Production Incentive of $1.08/kWh until June 30, 2020
  - *Estimated Annual Income:* $56,840 (production incentive); $4,128 (power sales)
  - *Estimated Annual Expenses:* $10,000
  - *Simple Payback:* 7.2 years

For more information: Linda Irvine, linda@nwseed.org, www.nwseed.org

Nonprofit Model

Nonprofits may engage with community shared solar projects in at least two ways: they may organize and administer a community shared solar project that shares benefits with participating members or they may solicit donations for a solar project. While this second option is not strictly "community shared solar," in that the donors do not share directly in the benefits of the solar installation, the donors do share indirectly, by lowering energy costs for their favored nonprofit and demonstrating environmental leadership. In addition, with emerging state policies such as virtual net metering and group billing, there may be possibilities for nonprofit project sponsors to share benefits with their donor/members. In a variation on nonprofit ownership, a nonprofit may partner with a third-party for-profit entity, which can own and install the system and take the tax benefits. This model has been deployed successfully in the California Multifamily Affordable Housing program and at other nonprofit locations throughout the country.[10]

> *If a nonprofit were to return some benefit to donors, (for example, a portion of production incentives or a share of electric savings) this would constitute a "quid pro quo" contribution and the donor could not deduct their entire contribution.*

Overview

Nonprofit organizations such as schools and churches are partnering with local citizens to develop community shared solar projects. Under this model, supporters of the nonprofit organization help finance the system through tax-deductible donations or direct investment in the project. The second option requires that the nonprofit comply with state and federal securities regulations. While the nonprofit is not eligible for the federal commercial ITC, it may be eligible for grants or other sources of foundation funding that would not otherwise be available to a business. An example of this model is the "Solar for Sakai" project on Bainbridge Island, Washington, in which a community nonprofit raised donations for a solar installation, and in turn, donated the installation to a local school.

Tax and Finance Issues for Nonprofit Projects

As non-taxpaying entities, nonprofit organizations typically are not eligible for tax incentives. However, donors to a nonprofit project can receive a tax benefit in the form of a tax deduction. The IRS allows taxpayers who itemize deductions to deduct verifiable charitable contributions made to qualified organizations. Of course, a tax deduction is much less valuable than a tax credit. For example, a $100 tax credit reduces taxes owed by $100 while a $100 tax deduction reduces taxes owed by $25 for a taxpayer in the 25% federal bracket.

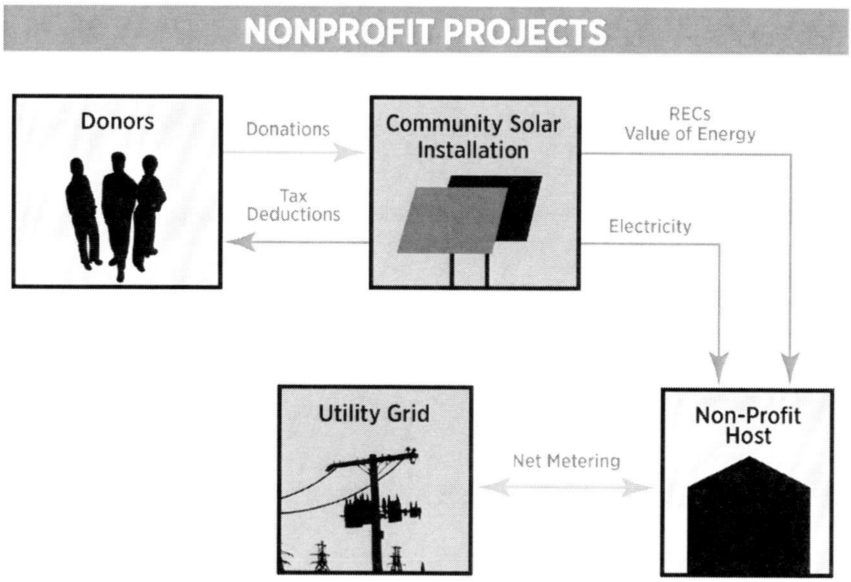

Donors can deduct their contributions to a community shared solar project if the project sponsor obtains tax-exempt status as a charitable organization under the Internal Revenue Code (26 U.S.C. § 501(c)(3)). Section 501(c)(3) organizations must be organized and operated exclusively for exempt purposes such as charitable, religious, educational, or scientific purposes. Section 501(c)(3) organizations may not be operated for the benefit of private interests and are restricted in how much time they can devote to lobbying activities. The Application for Recognition of Exemption under Section 501(c)(3) is IRS Form 1023.

Winthrop Community Solar Project, Washington

Following the 2010 launch of Okanogan County Electric Cooperative's (OCEC) first community shared solar project, co-op members who had been unable to participate were eager to develop another community shared solar project.

Project design and management was handled by Energy Solutions, who solicited the Town of Winthrop as project host and the Partnership for a Sustainable Methow (PSM) as project administrator. As a nonprofit with a mission to initiate, encourage, and support activities that foster long-term sustainability and wellbeing in the Methow Valley community, PSM was eligible for a Nonprofit Notification of Claim of Exemption from the Washington State Division of Securities. This exemption allowed PSM to offer ownership in the community shared solar project to members, contributors, or participants in the organization, or to relatives of community members. In early 2011, the opportunity to participate was announced through local press, radio, and the PSM website. Applications were processed on a first come, first served basis, ultimately attracting 49 investors to fully fund the community shared solar project in just six weeks. Investment levels ranged from $500 to $15,000, with investors participating at all levels.

Participating investors were not eligible to claim the 30% federal investment tax credit, which is unavailable to nonprofits and other entities that do not pay taxes. However, the high state production incentive for community shared solar projects using Washington-made materials partially made up for the loss of the tax credit. When the production incentive expires in June 2020, project ownership will be transferred to the Town of Winthrop.

- **Program Highlights**
    - *System Owner:* Participating OCEC members
    - *System Administrator:* Partnership for a Sustainable Methow
    - *System Host:* Town of Winthrop
    - *Installed Capacity:* 22.8-kW ground mounted array
    - *Participant Agreement:* Ownership purchased in $500 increments up to $15,000. Investors sign an ownership contract with PSM, which receives owners' investments, pays bills, and distributes production incentive to owners through June 2020. System ownership will then transfer to the project host, the Town of Winthrop.
    - *Electricity:* Net metering benefits accrue to Town of Winthrop (host), production incentive benefits accrue to participating OCEC members (owners)
    - *RECs:* Remain with participating OCEC members
    - *Number of Participants:* 49 investors

Photo from Ellen Lamiman, Energy Solutions.

- **Financial Details**
    - *Installed Cost:* \$200,000 or \$8.77/watt (cost to investors \$9.64/watt, includes insurance, bookkeeping, and administration costs)
    - *Capital Financing:* Project financed with owner investments, secured prior to construction
    - *Tax Credits:* None; federal tax credit cannot be claimed if project is not a business venture or is not placed on an owner's residential property.
    - *Grants:* None
    - *Rebates:* None
    - *Estimated Annual Payment to Participants:* \$72 per \$500 of investment
    - *Estimated ROI:* 30% by June 2020

For more information: Ellen Lamiman, elamiman@silicon-energy.com, (425) 320-6063, www.sustainablemethow.net

Solar for Sakai, Bainbridge Island, Washington

Community Energy Solutions, a nonprofit organization on Bainbridge Island, Washington, led the effort to raise funds for a solar installation at Sakai Intermediate School. Twenty-six community organizations or individuals made tax-deductible donations to Community Energy Solutions. The school owns the PV system and all of the resulting power and environmental attributes.

- **Program Highlights**
  - *System Owner:* Sakai Intermediate School
  - *Installed Capacity:* 5.1 kW
  - *Electricity:* Net metered
- **Financial Details**
  - *Installed Cost:* $50,000 or $9.80/watt (not including energy curriculum and monitoring)
  - *Grants:* $25,000 from utility (Puget Sound Energy)
  - *Donations:* $30,000 through Community Energy Solutions
  - *Production Incentive:* $0.15/kWh from state of WA

Summary of Benefit Allocation Options by Model

As evidenced by the examples above, there are many options for allocating the benefits of community shared solar within each sponsorship model. The following chart summarizes the most common options.

	Utility	**Special Purpose Entity**	**Nonprofit**
Electricity from Solar System	• Participants receive an estimated or actual kWh credit for their portion of project (virtual net metering) • Participants receive a monetary credit for the value of production for their portion of the project	• SPE sells the electricity to the utility (PPA) • SPE sells the electricity to the system host (SSA) • SPE assigns kWh to utility accounts per agreement with utility (virtual net metering) • Electricity from the system is netted against SPE members' group bill	• Nonprofit owner uses on-site and net meters • Nonprofit owner assigns to utility accounts per agreement with utility (virtual net metering) • Electricity from the system is netted against a group bill
Renewable Energy Credits	• Assigned to participants • Retired on participants' behalf • Retained by the utility	• Rights to RECs sold up front • RECs sold on an ongoing basis • Retained for participants	• Rights to RECs sold up front • RECs sold on an ongoing basis • Retained for nonprofit

	Utility	Special Purpose Entity	Nonprofit
Federal Tax Credits and Deductions	• Neither the commercial ITC nor the residential renewable energy tax credit is available to participants • If the utility has a tax appetite, it may use the commercial ITC • Normalization accounting rules will impact the value of the ITC for regulated utilities	• SPE can pass benefits of Commercial ITC through to participants • Only of use if participants have a tax appetite for passive income offsets	• Project donors can deduct the r donation on theitaxes • Nonprofits are not eligible for federal tax credits
Accelerated Depreciation (MACRS)	• Not available to participants • An investor-owned utility may be able to use MACRS, provided they own the system • To qualify for MACRS, regulated utilities must use normalization accounting	• SPE passes depreciation benefits through to the participants, subject to passive activity rules	• Not useful to nonprofits
State and Utility Rebates and Incentives	• Utility may qualify and use rebates/incentives to buy down the project costs; benefits are indirectly passed on to participants	• SPE may qualify and use rebates/incentives to buy down the project costs or pass through to participants	• Nonprofit may qualify and use rebates/incentives to buy down the project costs

SECTION 3: EMERGING STATE POLICIES TO SUPPORT COMMUNITY SHARED SOLAR

Over the last several years, a number of states have expanded their successful on-site solar programs by instituting policies that encourage innovative community shared solar programs. While each of these state programs varies considerably, a number of themes are emerging. For example, all of the current state-level programs require the solar array and the group members to be located within the same utility service territory. Other requirements to participate in "group" ownership benefits vary, but may include a cap on system size, proof of partial ownership, or limits on the type of ratepayers that can participate. Billing methods also vary; some programs offer one aggregate bill for the entire group, whereas others assign a pro-rated monetary credit on each member's bill.

State-level community shared solar policies can be grouped based on how the benefits of community shared solar are distributed. In general, there are three broad categories: group billing, virtual net metering, and joint ownership.

LOCAL FLAVOR

In Vermont, two well-known residents, Ben and Jerry, (the ice cream guys) decided to share the benefits of one solar installation on a shared electric bill. They hired AllEarthRenewables to build a solar array on Ben's guesthouse and informed their electric utility that the output of the installation should be netted against the combined consumption of both Ben's and Jerry's homes, in one bill. The solar panels offset all of the energy consumption at the guesthouse, and the remainder of the energy is applied toward offsetting the combined use of Ben and Jerry's homes. They get one electric bill, and split the offset 50/50. They don't have a formal contract, but it works because they are good pals with a long history of working together.

Group Billing

Group billing arrangements operate much like master metering in a multi-unit residential or commercial building. Under master metering, a landlord receives a single electric bill for all electricity usage within a building, including tenant load. The landlord then determines how to assign energy costs to individual tenants taking into account tenant leases. Group billing for community shared solar projects works in a similar way, except that participants do not need to reside in a single building. First, a utility produces a group bill showing all participants' energy consumption and relevant charges.

Then, output from a shared PV system is netted against the group bill. The remaining costs are allocated to participants according to an agreement between the participants. Under this framework, group billing allows multiple participants to receive net metering credits from a single renewable energy facility.

A drawback to group billing is that a customer representative must serve as a point of contact and an intermediary between a group of participants and a utility. The customer representative takes on tasks, such as billing and dispute resolution that expose the representative to administrative burdens. This framework may also raise concerns regarding the creditworthiness of a customer representative.

Vermont has expanded its net metering program to allow group billing for shared systems and this expansion has proven very popular.[11] In the service territories of Vermont's two largest utilities, Green Mountain Power and Central Vermont Public Service territory, over 22 groups have formed to share in the output of a renewable energy system with system sizes ranging from 1.5 to 199 kW. Vermont's program is not limited to solar energy systems. Any eligible renewable energy resource within Vermont's net metering program, including wind, small hydro, and biomethane can be installed under a group billing arrangement. In 2011, Vermont doubled the capacity limit for net metered systems, including group net metered systems, to 500 kW.[12]

Virtual Net Metering

Community shared renewables programs in Colorado, Delaware, Massachusetts, and California rely on virtual net metering to distribute economic benefits from a shared solar energy system. Similar to group billing, virtual net metering allows net metering credits generated by a single renewable system to offset load at multiple retail electric accounts within a utility's service territory. As with traditional net metering, credits appear on each individual customer's bill.

Colorado has implemented one of the most publicized and recognized community shared solar programs using virtual net metering, which it calls Community Solar Gardens. Colorado has allowed jointly owned systems (discussed below) for quite some time, although it has not formulated detailed program rules to support joint ownership. In 2010, Colorado authorized the Community Solar Gardens program under a subscription-based model.[13] In 2011, the Colorado Public Utilities Commission implemented rules governing the program.[14] The rules allow for substantial flexibility with regard to the structure of the Community Solar Garden entity. Regarding the virtual net metering component, the Community Solar Gardens program values a solar garden subscriber's bill credit according to the subscriber's "total aggregate retail rate," less a "reasonable charge" to account for the delivery, integration, and administration costs of the program. Stakeholders continue to discuss the calculation of the bill credit in another docket at the Commission (11A-418E) as part of the Commission's approval of Xcel Energy's 2012 Renewable Energy Standard Compliance Plan.

Unlike Colorado's program, Delaware's community shared program is open to any eligible renewable energy resource—solar, wind, ocean, geothermal, biogas, and small hydro—within Delaware's net metering program. Delaware passed the bill permitting the program in July 2010 and the Delaware Public Service Commission followed up with rules in June 2011.[15] Delaware's community shared renewables program allows community systems to be behind a customer's meter or off site. The value of the virtual net metering credit depends on whether or not a customer is on the same distribution feeder as the facility. If the customer is on the same distribution feeder as the facility, the credit is essentially valued at the customer's full retail rate. If it is not, the credit is essentially a generation-only credit. In other respects, Delaware's program structure is identical to the Community Renewables Model Program Rules, developed by the Interstate Renewable Energy Council (Appendix B).

Under Massachusetts' virtual net metering program, there are two avenues of participation:

1. A "neighborhood net metering" program allows neighborhood facilities to serve the energy needs of at least ten residential customers in a neighborhood group.
2. An alternative program allows participating net metered systems to allocate monthly excess generation to one or more customers within a distribution company's service territory.

Under Massachusetts' neighborhood net metering program, a renewable energy system must be behind a participating customer's meter. However, only a minimal amount of load needs to be present on site. In fact, even "parasitic" load needed to run a facility is allowed to count toward meeting on-site load requirements. Kilowatt-hour credits generated by a renewable energy system are allocated to participating customer accounts by the participating utility. Utilities are not required to include the distribution component of participants' applicable retail rate within neighborhood net metering credits.

Under an alternative program, and contrasting what is typically seen in net metering, Massachusetts allows any customer with a net-metered system to allocate credits associated with monthly excess generation from a system to other customers of the same distribution company. Customers designated by the owner of the net-metered system receive a net metering credit that reflects the host customer's fully bundled retail rate. The net metering credit offered to designated customers is calculated using the retail rate of the host customer (cents per kWh), multiplied by the allocation of kWh for the designated customer. While on-site load must be present where the net-metered system is installed, as with neighborhood net metering rules, parasitic load qualifies as on-site load. This alternative program is very flexible in who can participate and offers a more financially attractive net metering credit than the neighborhood net metering program.

SOLAR FOR ALL

The nonprofit San Diego Community Housing Corporation (SDCHC) partnered with a third party, Everyday Energy, to put a 20-kW system on its Hacienda Townhomes property. Everyday Energy installed and owns the system on the 52-unit apartment building, taking advantage of the tax benefits that are not available to the nonprofit Housing Corp. SDCHC signed a 20-year solar services agreement with Everyday Energy under which they will pay a flat fee to cover maintenance and electric services from the installation. An electric meter measures the energy flow directly to the grid, and the utility (San Diego Gas & Electric) credits the tenants and common areas as directed in the Virtual Net Metering agreement. Residents will save a projected 30% on their electric bills.

The California Public Utilities Commission recently expanded the availability of virtual net metering in California to all multitenant buildings in the state. Up until this expansion, under California's Multifamily Affordable Solar Housing (MASH) program, residents of multifamily, low-income complexes, such as the SDCHC townhomes in San Diego, have been allowed to receive bill credits from a single on-site PV system.[16] The building owner allocates net metering credits to individual tenants and a building's common load. Virtual net metering allows the building owner to avoid building a separate solar energy system with a separate inverter for each tenant, which saves considerable funds. According to a program report issued in August 2011, 325 projects are eligible for participation in the MASH program representing over 20 MW of capacity.[17] In July 2011, the Commission expanded the types of customers eligible for virtual net metering to tenants in any multi-tenant or multi-meter property—not just affordable housing properties. In addition, the Commission allowed for virtual net metering credits to be shared throughout an entire affordable housing property, as long as that property is on contiguous parcels and under common ownership.[18] This change opens virtual net metering to a much broader group of customers and signals a significant expansion in California's net metering program.

Joint Ownership

Following the precedent set by successful community shared wind programs, a few states are exploring options for distributing the benefits of participating in a community shared renewable energy program through frameworks similar to wholesale power sale arrangements. The community shared wind movement was motivated, in part, by a desire to promote rural development through expanded citizen investment opportunities. By allowing citizens to "piggyback" their projects onto larger wind projects, communities could benefit from economies of scale. This history led to a primary difference between the emergence of community shared solar and wind: Community shared wind uses a technology that began as utility-scale application and moved into smaller scale applications. In contrast, community shared solar uses a technology that began with on-site systems and applies it to larger solutions.

Maine's Community-Based Renewable Energy Pilot Program law[19] allows "locally owned electricity generating facilities" with at least 51% ownership by "qualifying local owners" to elect one of two incentive mechanisms. Under the first, qualifying local owners can enter into a long-term contract to sell output from a facility to a transmission and distribution utility. The contract price for energy may vary over the course of a year, but the average price, weighted based on the expected output of a facility, may not exceed $0.10 per kWh. This price only includes the value of a power sale and does not include a purchase of RECs. A significant downside of this approach is that a payment for power sales to a wholesale or retail purchaser results in taxable income at a federal level and possibly at a state level. Depending on the tax bracket a particular customer faces, the taxation of payments for power sales can significantly decrease the size of benefits available to participating customers.

Under Maine's second incentive option, generation is virtually net metered to joint owners in proportion to the owners' stake in a system. For example, a 50% owner would receive 50% of the net metering credits generated by a system though virtual net metering.

Washington's community solar rules allow for ownership of community shared solar projects up to 75 kW that are either jointly owned by individuals, businesses, and nonprofits or owned by a utility and voluntarily funded by the utility's ratepayers. Participants receive production incentives based on their proportional share of the output of a project. In addition, in the case of utility-owned projects, participants receive the value of the electricity. Washington's community solar incentives are among the most generous in the world if projects use inverters and modules made in Washington. For such systems, the production incentive is set at $1.08 per kWh through June 2020, but is subject to dilution if incentive payments exceed 0.5% of utility gross revenue in a given year.

SECTION 4: TAX POLICIES AND INCENTIVES

Federal tax incentives for solar systems are especially valuable and tend to be a primary driver in the design of project structures and financing strategies. This section introduces some state and federal tax policies that impact community shared solar projects, as well as other federal financial incentives in the form of grants, bonds, and loans. For details on tax issues specific to each ownership model, see Section 2, Community Shared Solar Project Models.

Federal tax incentives provide significant support to solar projects, offsetting approximately 56% of the installed cost of a commercially owned PV system and 30% of a residential installation.[20] However, community shared solar project designers should be aware that federal tax incentives were developed with either individually owned residential installations or comercially owned projects in mind. Community-scale projects do not fit squarely into either category, which makes it challenging to design projects that can make use of either the residential or commercial tax credits. For example, the residential Renewable Energy Tax Credit is not available to community shared solar projects because it only applies to taxpayers who install a solar system on their own residences.

Receiving any kind of financial benefit or loss from participation in a community solar project could have tax consequences for the participant. In addition, tax incentives can interact in complicated ways, and project organizers should seek professional advice before including tax incentives in a project plan.

Proposed legislation at the federal level could make it easier to use tax credits for community shared solar. Senator Mark Udall (CO) proposed the SUN Act 2011, which would allow individuals to claim the residential tax credit when purchasing solar panels in a community shared solar project. For more information and updates, see Senator Udall's website: www.markudall.senate.gov/.

Tax incentives vary widely, depending on the status of the project sponsor. For example, investor-owned utilities are eligible for tax incentives that are unavailable to municipal utilities or electric cooperatives. Nonprofit projects cannot use solar tax benefits, but donations to them are tax-deductible. Special Purpose Entity business projects have the greatest flexibility for taking advantage of federal tax incentives. As a result, a host of project business structures—some of which are very complicated and require significant legal expertise—have been created to maximize federal tax incentives. These structures are discussed in greater detail in Section 2, Community Shared Solar Project Models.

The following federal incentives may be applicable to a community shared solar installation, depending on the details of each project. Additional detail on each of these federal incentives can be found on the Database of State Incentives for Renewables & Efficiency (DSIRE) located at www.dsireusa.org/.

Business Energy Investment Tax Credit ("Commercial ITC")

The Commercial ITC is one of the most valuable incentives available for solar energy. The Commercial ITC allows commercial, industrial, and non-public utility owners of PV systems to take a one-time tax credit equivalent to 30% of qualified installed costs. Under the Commercial ITC, the owner of the PV system for tax purposes can be different from the owner of the host property. Therefore, the use of a third party to finance systems has emerged as a leading trend in the solar industry. The tax credit can be used to offset regular tax and alternative minimum tax (AMT). The Commercial ITC is currently available for systems that are placed in service before the end of 2016. There is no cap on the amount of the Commercial ITC. Unused credits can be carried forward for up to 20 years. Commercial entities will likely pay income taxes on any up-front rebate or cash incentive the entities receive. In this case, entities not have to reduce the "cost basis" by the amount of the rebate before calculating the Commercial ITC. After January 1, 2017, owners of qualifying solar facilities will be eligible to claim a 10% ITC.

Eligibility and timing issues are complex. For a discussion of these issues, as well as the basis reduction and allocation issues, see the DSIRE website: www.dsireusa.org/solar/incentives.

The Commercial ITC is available to private utilities and SPEs owing federal taxes.

The American Recovery and Reinvestment Act of 2009 created a cash grant alternative to the Commercial ITC. The owner of a qualified solar facility eligible for the ITC could instead elect to receive a grant for approximately the same value. This was especially valuable to taxpaying entities that could not take full advantage of the ITC due to lack of tax appetite. The Section 1603 Treasury Grant expired in 2011.

Modified Accelerated Cost Recovery System (MACRS)

In addition to grants and tax credits, federal tax policy allows businesses (but not individuals) to depreciate their investments in solar projects on an accelerated basis. "Depreciation" refers to the concept that over time, assets such as equipment lose value and will eventually need to be replaced. To account for this reduction in asset value, businesses record an expense over a set period of time. For qualified solar projects, this period is five years. Subject to certain restrictions, an owner with other sources of passive income can offset that income with losses generated by accelerated depreciation deductions under the modified accelerated cost recovery system (MACRS).

For projects placed in service by the end of 2012, bonus depreciation is available. This allows the owner to deduct 50% of the adjusted basis of an eligible solar system in the first year. For projects taking the ITC, the depreciable basis must be reduced by half the value of the ITC. For example, if the ITC equals 30% of project costs, the depreciable basis is reduced by 15%.

The IRS publishes schedules that detail how different asset classes should be depreciated. For additional information, please consult IRS Publication #946. A more detailed discussion of using tax benefits is provided in Section 2, in the discussion of the Special Purpose Entity ownership model.

Tax Credit Bonds

Qualified tax credit bonds are a mechanism to lower the cost of debt financing for non-taxpaying entities such as government agencies, municipal utilities and electric cooperatives. Two tax credit bonds in particular—Clean Renewable Energy Bonds (CREBs) and Qualified Energy Conservation Bonds (QECBs)—were created to finance renewable energy projects and programs. All available tax credits have been awarded and no additional funding is expected. However, project organizers may find that some awardees have unallocated funds that might be used for a community shared solar project.

Clean Renewable Energy Bonds (CREBS)

CREBs are tax credit bonds that can be used by government entities, municipal utilities and electric cooperatives to finance solar installations and other renewable energy projects. Ashland, Oregon used the proceeds from a CREB to partially finance its Solar Pioneers II community shared solar project in 2008.

Qualified Energy Conservation Bonds (QECBS)

QECBs are tax credit bonds similar to CREBs. The advantage of QECBs is that in addition to using them to finance renewable energy projects, QECBs can also be issued for energy efficiency projects and green community programs, among other things. In addition, up to 30% of a QECB allocation can be used for private sector activities. To date, the authors of this Guide are unaware of a community shared solar project that has used QECBs.

Federal Grants

While not necessarily a source of long-term funding, federal grants can be used to reduce the cost of a community shared solar project. Such grants could lower the cost of the PV system installation or subsidize the cost of participation in a community shared solar project. In 2009 and 2010, enhanced funding was provided for State Energy Programs and Energy Efficiency and Conservation Block Grant Programs (EECBG). For rural communities, there may be USDA grants and loans available through the Rural Energy for America Program (REAP).

SolarShare Community Solar Bonds

In Ontario, Canada, nonprofit cooperative SolarShare has introduced a new way to engage communities in solar financing. Using their co-op model, SolarShare allows Ontario residents and businesses to benefit from investing in bonds backed by large commercial and smaller rural solar projects. Co-op membership costs $40, and each member is able to invest $1000 in bonds. Bond investments are used to finance fully completed projects, which shields bondholders from pre-development and construction risks. Each project is backed with a 20-year PPA with Ontario Power Authority with fixed pricing for the power produced, ensuring a steady and long-term revenue stream. Bond repayments are made semiannually with 5% interest, and are fully repayable after completion of a five-year term. Currently, the SolarShare project portfolio consists of 18 solar installations. For more information, see www.solarbonds.ca.

Examples of projects benefiting from federal grant funding include Seattle City Light's Community Solar Initiative funded under DOE's Solar Energy Technologies Program, the second phase of St. George, Utah's SunSmart Community Solar program using Recovery Act funding, and APS's Community Power Project using a High Penetration Solar Deployment grant from the DOE's SunShot Initiative.

State and Local Tax Considerations

Tax issues vary considerably from state to state and among localities. However, there are several common issues that project developers should consider when planning and structuring their projects. Taxes in any of the categories below could impose a significant cost on the project. Project developers should determine which taxes will apply to their project and who will be responsible for the cost. Taxation issues can become especially complex when a project involves both taxable and tax-exempt entities. Considerations include the following:

Net Income Tax: Most states impose a net income tax modeled on the federal system. Thus, any revenue generated by a project will likely be subject to both state and federal income taxes. Some states offer investment tax credits that can be taken in addition to the federal Commercial ITC or other income tax credits and deductions for renewable energy. In Utah, for example, the State's residential income tax credit is available to participants in community shared solar projects owned by qualifying entities (municipalities, counties, etc.), such as the SunSmart program in St. George.[21]

Sales and Use Taxes: Most states impose a sales tax on sales of tangible personal property. Some states also impose a use tax on sales of certain services or a transfer tax on sales of real property. For solar facilities, most state sales taxes will apply to the purchase of solar equipment, but usually not to the sale and use of electricity. Many states offer sales tax incentives for solar facilities in the form of reduced rates, exemptions or rebates.

Property Tax: Nearly all states impose a property tax that is assessed annually, based on the value of real property. Most states also tax tangible personal property that is used for business purposes. For property tax purposes, assessment values might be determined by a central state authority or by a local assessor's office. As with sales taxes, many states offer property tax incentives for solar facilities in the form of exemptions or special assessments.

Excise Taxes: Some states and municipalities impose excise taxes that could potentially apply to a solar facility. An excise tax is special tax imposed on particular goods or activities, such as a gasoline tax or gambling tax.

Interactions among State and Federal Incentives

The Commercial ITC is valued at 30% of the tax basis of thc solar facility. The "basis" typically means the cost of buying and installing the facility. But certain factors can reduce the basis from which the 30% is taken. Other financial incentives (such as state rebates and grants) will reduce the taxpayer's basis for calculating the ITC, unless the incentives are considered taxable income to the taxpayer. If the incentive is considered taxable income, it does not need to be subtracted from the cost basis. These rules prevent "double-dipping" that would come from receiving both a tax-free incentive and a tax credit.

SECTION 5: SECURITIES COMPLIANCE

Community shared solar projects can be structured to create ownership models that monetize financial incentives, capitalize on favorable government and utility policies, and expand ownership opportunities. When devising a creative business model, though, the project organizer should consider whether the model involves the issuance of securities, and, if so, what federal and state securities laws may be involved. A full review of state and federal securities requirements related to small offerings is beyond the scope of this guide, but this discussion is intended to offer a foundation for project organizers to research the issue.

Any entity, no matter how small, that attempts to raise capital may be deemed to be issuing securities if it offers or sells stock, membership units, partnership interests or other types of participation interests. If the project is deemed to be offering a security, the project will incur substantially more time and expense in ensuring that it complies with the applicable state and federal securities laws. The consequences of failing to comply can be severe and the project, its directors, officers, and employees involved in the offer and sale of the security may be subject to liability for such failure.

The securities laws are intended to protect individuals who provide financial support for a project with an expectation to receive profits from the efforts of others, or with the expectation to receive a valuable benefit when the investor does not have control over the managerial decisions of the venture. Compliance with securities laws requires registering the offering with the Securities Exchange Commission (SEC) and the applicable state regulatory agency or finding a specifically-defined state and federal exemption from the registration requirements. Most states' securities laws have parallels to the federal requirements, but many states require additional filings, even if their exemptions are similar in substance to the federal exemptions.

Registration can be a time-consuming and expensive process that includes filing a formal registration statement with the SEC and preparing extensive disclosure documents called an "offering memorandum." However, even with a registration exemption, filings and the preparation of offering documents may still be required, depending on the participants in the project. Many projects will not be able to support the up-front costs of securities compliance.

The definitions of a "security" under federal and state laws include a long list of financial instruments and agreements. Federal and various state definitions are not identical, but commonly include, any note, stock, bond, evidence of indebtedness, certificate of interest or participation in any profit sharing agreement, or investment contract.

SOLAR CROWDFUNDING

Crowdfunding is a cooperative financing approach that occurs when many small investments are aggregated to collectively finance a single initiative. In November 2011, the House of Representatives overwhelmingly passed the Entrepreneur Access to Capital Act (HR 2930), which would amend the current regulatory landscape by allowing startups to offer and sell securities through crowdfunding and social networking websites. This Act would allow groups of individuals to collectively invest in and own solar systems without having to submit to the SEC's lengthy registration and reporting requirements. One example of solar crowdfunding in action can be found in Solar Mosaic (www.solarmosaic.com), a Berkeley-based company that connects individual investors with solar projects hosted on community sites. The site host leases the solar system from the investors, and investors are paid back over time by monthly lease payments (and other available incentives) processed through Solar Mosaic.

A common exemption used by smaller-scale non-utility-owned projects is the private placement exemption, which allows a company to raise investment capital from a certain number of investors. All private placement exemptions limit the number of individuals or entities to whom the securities can be offered. The level of the disclosure requirements is determined according to the net worth or income level of the investor and/or the relationship of the investor to the entity issuing the security (for example, if the investor is acting as the executive officer or director of the entity).

The most relevant test for analyzing whether a contract or an investment is a security under federal law is the "Investment Contract Test." Many states have additional criteria for determining the existence of a security, but the basic components are similar to the Investment Contract Test. A security exists if (1) a person invests money or property, (2) in a common enterprise (i.e., an enterprise in which the benefit to the investor is dependent on the participation of others), (3) with an expectation of profits or other valuable benefits, (4) solely or primarily from the efforts of someone other than the person providing the money or, in other words, without the right to exercise practical and actual control over the managerial decisions of the enterprise.

It follows that the terminology used to describe participation in a community shared solar project should avoid references to "shares" or "stock," as those are the classic terms used to describe securities issued by a corporation and might create an expectation of profits and other rights customarily associated with stock or shares. All marketing and promotional materials used for the project should refrain from making any statements suggesting that an investment or other opportunity to make money is being offered to participants. However, regardless of the label applied by the project sponsors, there is always some risk that the securities regulators or a court will deem the participation to be a security.

In a utility-owned model, in which the utility enters into a contract or arrangement with its retail customer to provide electricity generated by a project, there is a risk that the contract or arrangement could be deemed a security if the customer is required to finance a part of the project and if the customer has an expectation of getting some kind of profit over and above the value of the electricity it receives.

To the extent that a retail customer agrees to purchase solar power from a utility and to pay a specified, generally applicable rate for the solar power used and the customer is billed periodically based on recent past use, just like the arrangements for purchasing other power, it is less likely that the customer would be viewed as making an investment of money in the project. In contrast, if the customer is required to buy a panel or make payments in excess of the retail market rate for the solar power, it is more likely that the customer will be viewed as making an investment of money. Therefore, the utility must take care to ensure that the rate charged for the solar power does not contain a charge for the customer's acquisition of an interest in the project or a panel. In addition, a payment is more likely to be an investment if the customer pays an up-front amount in return for an undetermined amount of solar power over a period of time that may also be undetermined.

In order to reduce the likelihood that the contract is a security, payments made under the contract could be: (1) applicable to a specific, relatively short period of time (e.g., monthly, quarterly); (2) due after solar power is provided;

and (3) according to a specified, generally applicable market rate per unit that does not include a component for the purchase by the customer of an interest in the project. To the extent possible, the contract, pricing and billing arrangements, and related materials should resemble a customary consumer purchase of non-solar electricity and should not be marketed to emphasize that the amount of solar power sold to customers depends on the participation of other customers or the success of the utility in obtaining subscribing customers or in operating the project. The corollary is that customer dollars cannot be used up front to finance the project.

SECTION 6: GETTING STARTED

There are many legal, financial, and project design considerations to address before launching a successful community shared solar project. This section outlines a general process to help community organizers and project developers move concepts to completion efficiently. In addition to consulting this guide, project developers are encouraged to consult The Resource Innovation Group's online Community Solar Tool for help developing community-owned projects (see http://communitysolar.dyndns.org). Although this decision support tool was developed for the University of Oregon and contains state-specific information, it poses questions relevant to all community shared solar projects and is undergoing customization on a state-by-state basis as funding permits.

> *It took us over two years to develop our project structure and only two months to find our members. –David Brosch, University Park Community Solar*

Like many construction projects, community shared solar project development projects can be broken down into phases including: **feasibility, project development, construction, operations and maintenance, and decommissioning.** It is important to note that phases can often overlap and are not necessarily completed in the order listed.

Feasibility Analysis Phase

The first step is to conduct a comprehensive feasibility analysis. This analysis should determine if there is a good project site with an adequate solar resource, identify a project team and supporters, prepare an initial financing plan, confirm absence of major obstacles, and gauge the local community and utility's receptivity to a project.

Project Development Phase

If the feasibility analysis is positive, the project can move to the development phase. At this point, it may be helpful to document the project details in a business plan (which may be required to secure financing) or project charter.

Site Selection and Resource Evaluation

Proper siting includes a site analysis for any potential shading, as well as determining optimal tilt of the modules, location of inverters and other system components, wiring distances, foundation or structural support, and security or public access requirements. Project owners must also obtain exclusive rights to build the solar project if they are not the property owners. This is usually negotiated through a land lease agreement with the property owner and site host. Careful consideration should be given to site selection to minimize the environmental footprint and harmonize with existing land uses.

Understanding the amount of solar resource and the effects of climate and latitude on solar energy production is critical to finalizing the system location and obtaining estimates for financial modeling. Typically, project organizers will rely on solar resource maps or solar energy production calculators, such as PV Watts or RETScreen, to get an initial assessment of the solar resource.

Financing

To obtain financing for a project, a financial pro forma must be created to model the proposed system's costs, revenue (from the production estimates), and the interaction of incentives and financing. This document reveals the financial viability of the project, and is required for any project proposal. A very basic sample budget is provided after this discussion to suggest the broad categories of expenses and income that should be considered.

Ownership Structure

The ownership structure of the project will need to be finalized and the business model chosen. The project owner(s) may also need to consult legal and tax professionals to ensure the entity is properly structured to minimize risks to the site host, investors, and participants.

Permitting and Environmental Review

The permitting process for a community shared solar project depends on the location, size, and type of project. At minimum, the project will require an electrical permit. A building permit is often necessary, especially if the PV array is a stand-alone structure. The best course of action is to check with the local planning department early on, as the permit and environmental compliance requirements may influence the design and siting of the project.

Interconnection and Power Arrangement

The local utility will be involved in interconnecting the system to the electric grid. Utilities generally follow a standard interconnection process and have agreements that must be completed prior to construction. In addition to connecting the system to the distribution system, the agreement must account for arrangements for transferring the power "benefits." This is usually negotiated through a power sales agreement between the project owner and the utility or host in the form of a PPA, SSA, net metering, or other contractual arrangement.

Procurement and Contracting

For this type of project, it is common to issue a request for proposals (RFP). The RFP can be fairly broad, allowing solar professionals to offer their recommended system design and specifications, or fairly specific, to compare bids on pre-determined project specifications. After identifying solar professionals or receiving proposals in response to an RFP, it is important for project owners to evaluate them as they would evaluate other types of installers and contractors. Professional credentials are one indication of a PV installer's knowledge and qualifications. The North American Board of Certified Energy Practitioners (NABCEP) offers a well-respected voluntary certification program for PV installers.

Construction Phase

Choosing a solar contractor and construction manager is an important decision. In recent years, it has become easier to locate and contact those in the solar field. Tools available to help identify local professionals include www.findsolar.com and the national Solar Energy Industries Association (SEIA.org).

Operations and Maintenance Phase

Operating a community shared solar project requires ongoing record keeping and timely filing of paperwork. Among other things, a project administrator may have to file tax forms and business license renewals, distribute incentive payments, sell RECs, and keep the insurance, lease and other payments up to date.

Maintenance, though fairly simple for PV systems, is essential to long-term management of a community shared solar system. Modules may need to be cleaned, but more importantly, meters and inverters need to be monitored to ensure that the system is operating as expected. Various monitoring systems are available, offering options from instant email alerts when an inverter malfunctions to online daily performance monitoring. A good monitoring system enables a system manager to minimize down time, protecting the participants' investment. The project budget should include funds for monitoring, ongoing maintenance costs, and parts replacement. In particular, it is helpful to include a reserve fund for future inverter replacement.

Decommissioning or Exit Strategy

Although solar panels could easily last 25 years or longer, every project must consider the ultimate disposition of the solar installation. Whether the plan is to sell the project to the host, renew a lease, or remove the panels, a solid project plan has defined the options for exiting from the community shared solar project and potentially restoring the site to its original condition.

Community Shared Solar Project: Sample Budget

The following budget template provides sample categories for a typical community shared solar project budget.

Note that the budget does not include the cost of labor to organize and develop the project, which could amount to a one or more years of full-time work. Depending on how the project is developed (by a utility, an SPE or a nonprofit), the developer role could be volunteer or paid.

SITE DEVELOPMENT COSTS	
Design	$
Permits	$
Electrical/Meter Upgrades	$
Fencing/Security	$
Educational Kiosk	$
PROJECT DEVELOPMENT COSTS	
Consulting	$
Legal	$
RFP	$
SYSTEM COSTS	
PV Panels	$
Inverters	$
Ground Mount/Racking System	$
Balance of System Costs	$
TOTAL INSTALLED COST	$
MINUS GRANTS AND REBATES	
1603 Treasury Grant	$
Commercial ITC	$
Other Grants and Rebates	$
NET INSTALLED COST	$
ANNUAL OPERATING EXPENSES	
Bookkeeping	$
Accounting	$
Legal	$
System Monitoring	$
Insurance	$
Lease	$
Sinking Fund: Inverter Replacement	$
Taxes	$
TOTAL ANNUAL OPERATING EXPENSES	$
ANNUAL INCOME	
Sale of Electricity	$
Sale of RECs	$
Production Incentive, if Available	$

Community Shared Solar Project Development Worksheet

The following worksheet suggests many steps involved in organizing a project, but it is not comprehensive. Project organizers will need to create their own list of steps, based on their unique circumstances.

FEASIBILITY ANALYSIS
Assess site for solar access
Secure control of property and/or site
Evaluate the solar resource
Understand participant motivation
Conduct market research/focus groups/surveys
Investigate interconnection options
Research financing mechanisms
Gauge community receptivity and support
PROJECT DEVELOPMENT
Prepare a financial plan
Determine ownership structure
Develop operating agreement between host and project owner (if different)
Develop participant agreement
Obtain legal and tax consultation for contracts
Define system and other technical specifications
Execute agreement for the sale of power
Complete permitting and environmental compliance requirements
Execute interconnection agreement
Conduct an RFP for design/build
CONSTRUCTION
Prepare the site for construction: grading, road improvements, other
Dig trenches, lay cables, install transformer(s)
Install fencing and site security features
Complete inspections and commissioning
Restore site/surrounding vegetation
Complete paperwork for incentives
OPERATIONS & MAINTENANCE
Schedule and perform panel cleaning
Save for inverter replacement
Monitor system output
Distribute benefits to participants (incentives, tax credits, etc.)
File tax returns, state production incentive paperwork
File annual business license requirements

Section 7: Resources

Communities interested in implementing a solar project will need a more thorough understanding of many of the topics in this guide. The resources listed in this section can provide much of that information.

Organizations and Institutions

- Through DOE's SunShot Initiative, local governments are working to accelerate the adoption of solar energy technologies for a cleaner, more secure energy future. The website offers case studies, policy updates, and news of solar activities across the country. www4.eere.energy.gov/solar/sunshot/ resource_center/
- The Database of State Incentives for Renewables and Efficiency (DSIRE) is a comprehensive source of information on state, local, utility, and federal incentives that promote renewable energy and energy efficiency. www.dsireusa.org
- The Office of Energy Efficiency and Renewable Energy (EERE) works to strengthen the United States' energy security, environmental quality, and economic vitality in public-private partnerships. www.eere.energy.gov
- USDA Rural Development provides funding for the development and commercialization of renewable energy technologies in rural communities. The Rural Energy for America Program (REAP) offers grants and loans to help small rural businesses deploy renewable energy projects. www.rurdev.usda.gov/rd/energy
- The Bonneville Environmental Foundation (BEF) supports the development of renewable energy and watershed restoration and empowers people to shrink their carbon footprints. BEF's Project Management Group assists with the funding and construction of solar installations in communities throughout the Northwest. www.b-e-f.org
- Northwest Sustainable Energy for Economic Development (North-west SEED) empowers community-scale clean energy through targeted technical assistance, education and outreach. Northwest SEED seeks to increase responsible use of clean, renewable energy with maximum local control by providing on-the ground support to

communities in planning and implementing clean energy projects. www.nwseed.org/

- The American Solar Energy Society (ASES) is a nonprofit organization dedicated to increasing the use of solar energy, energy efficiency, and other sustainable technologies in the United States. This website is a good source for information about solar technology and professionals. www.ases.org/
- The Interstate Renewable Energy Council (IREC) is a nonprofit membership-based organization that provides a national forum in which public and private organizations involved with renewable energy may gather, disseminate and exchange information and engage in cooperative efforts. IREC's website offers the latest policy and practical solutions for tough renewable energy issues. www.irecusa.org/
- The Vote Solar Initiative works at the state, federal and local level to implement programs and policies that allow strong solar markets to grow. www.votesolar.org/

Publications and Online Tools

The Online Community Solar Tool, University of Oregon and The Resource innovation Group, is an online decision tool that provides a framework for making program development and design decisions. http://communitysolar.dyndns.org/

The Community Power Network offers examples and inspiration for community scale projects across the United States. The site includes a wiki to learn and share from other projects. www.communitypowernetwork.com/

Solar Resource Guide: An Overview for Congregations, California Interfaith Power & Light Network, July 2011. http://interfaithpower.org/resources/solar-resource-guide

Solar Powering Your Community: A Guide for Local Governments, U.S. Department of Energy (DOE), 201, includes case studies and lessons learned from Solar America Communities. www.solaramericacommunities.energy.gov/resources/guide_for_local_governments.

Community Solar Power: Obstacles and Opportunities, Institute for Local Self-Reliance, September 2010, profiles community shared solar projects, the policies that enabled them, and the barriers that remain. www.ilsr.org/

Financing Non-Residential Photovoltaic Projects: Options and Implications, Lawrence Berkeley National Laboratory, January 2009, examines the role of financial innovation in PV market penetration. This report looks at how financing structures currently being used to support nonresidential PV deployment have emerged as a way to extract the most value from a patchwork of federal and state policy initiatives. http://eetd.lbl.gov/ea/ems/reports/lbnl-1410e.pdf

Lex Helius: the Law of Solar Energy (3rd Edition), Stoel Rives, 2009 (See especially, Chapter 7: Financing) www.stoel.com/showarticle.aspx?Show=2886

APPENDIX A

Business Formation and Types: Special Project Entities for Community Shared Solar Projects

Below are descriptions of the primary business entities suitable for community shared solar projects, the key characteristics, and the major advantages and disadvantages each entity might have. Note: Characteristics commonly attributable to these business entities are discussed, but legal requirements can vary from state to state. State law may also establish default rules that can be changed by agreement among the business owners.

General Partnerships

A general partnership is an association of two or more people working together in a common business enterprise. There are few formal requirements for establishing a partnership and if the partners fail to enter into a written partnership agreement, the default provisions of the state partnership laws will govern the relationship of the partners. However, most partners choose to enter into a written agreement.

Advantages and Disadvantages of Forming as a General Partnership

The key advantage of organizing as a general partnership is the ease of formation and the flexibility in the relationship between the partners. General partnerships require little, if any, paperwork for formation or operation. General partnerships also allow for "pass-through" taxation, instead of the "double" taxation that may be applied to corporations. Additionally, most general partnership interests will not be treated as securities because all the partners contribute equally to the decision making processes and participate in management of the business. General partnerships, however, have several key disadvantages. First, and most important, each partner is individually liable for the partnership debts. This means that if the partnership cannot pay its debts, the creditors can look to the individual partners to satisfy those debts. Because of the lack of limited liability, general partnerships have fallen in popularity as a business entity in recent years. Second, the preparation of a partnership agreement requires the assistance of legal counsel and can be expensive, depending on the complexity of the partners' relationships. Third, because of the close personal relationships inherent in a general partnership, partnership interests cannot usually be easily transferred or sold. Unless a partnership agreement so provides, it can be challenging to admit new or substitute partners.

Formalities

As discussed above, in theory there are few, if any, formal requirements for forming general partnerships. Similarly, there usually are few requirements for operation, but states usually establish some default rules to govern if partners do not enter their own agreements. For example, in the absence of an agreement otherwise, the default rules usually provide that partners have equal control over businesses and equal share in profits and losses. Partnerships are "pass-through" entities, which means that profits and losses pass through to individual partners. The partnership is not a separate taxpaying entity; rather, the partners report profits and losses from the partnership on their individual tax returns.

Limited Partnerships

A limited partnership is a business entity comprised of two or more partners who operate or manage a business together. In every limited partnership, there are two types of partners: general partners and limited partners. The general partner usually invests significantly less capital than the limited partner(s) and has a significantly smaller ownership stake. Unlike general partnerships, limited partnerships have the ability to limit both the liability risk and the business involvement of certain partners known as "limited partners," but the general partner has unlimited liability. This feature is particularly useful for attracting "passive" investment partners who would like to participate in the profits of the business, but not necessarily take on its risks or daily operations.

General partners manage the company's day-to-day operations and are liable for the debts of the partnership. Because they are responsible for any debts or lawsuits incurred by the partnership, general partners often form limited liability entities, such as corporations or LLCs (both discussed below), to protect themselves from liability.

Limited partners contribute capital to the partnership but do not (and generally cannot) participate in the daily operations of the company. As an added benefit, they are also shielded from company debts and other liabilities. Limited partnerships are a popular choice for individuals who lack the time or expertise to run a business, but would like to share in the profits.

Advantages and Disadvantages of Forming as a Limited Partnership

There are several advantages to the limited partnership entity. The limited partners have limited liability and the limited partnership interests may be able to be sold easily without dissolving the limited partnership as an entity. The option of being a limited partner can attract investors because the investors' liability is limited. However, with certain exceptions, the limited partners have to refrain from dabbling in management; if a limited partner becomes too involved in the partnership's daily operations, the limited partner's status could be altered to that of a general partner, with the attendant loss of limited liability.

While limited partnerships are relatively easy to form, a limited partnership agreement is essential to govern the relationships of the parties, especially the contribution of additional capital and the allocation of profits and losses.

The major disadvantages of the limited partnership are first, that the general partner of a partnership assumes personal liability for the partnership's obligations and debts, and second, the passive nature of the limited partner's involvement carries the likelihood that the limited partnership interest will be deemed to be a security.

Formalities

Most states impose more requirements for forming a limited partnership than for a general partnership, such as filing a certificate of formation.

Limited Liability Companies (LLCs)

A limited liability company, usually called an LLC, is a separate and distinct legal entity. An LLC provides the limited liability protection for its owners (known as members) with the pass-through benefits and flexibility of a partnership. The members of an LLC are not personally liable for its debts and liabilities, and also have the benefit of being taxed only once on their profits.

Because LLCs have only been around for about 30 years, smaller banks may be reluctant to extend credit to LLCs. Further, with such a short history, many legal issues that arise in connection with the LLC format have not been settled.

An LLC may be managed by either (1) the members or (2) one or more managers. If a limited liability company is managed by the members, then the owners are directly responsible for running the company (a "member-managed" LLC). A "manager" is a person elected by the members to manage the LLC. In this context, a manager is similar to a director of a corporation. A manager can be, but is not required to be, a member. If an LLC is managed by managers, then its members are not directly responsible for running the company and the passive nature of a non-managing member's involvement makes it likely that the membership interest will be deemed to be a security.

LLC ownership can be expressed in two ways: (1) by percentage; and (2) by membership units, which are similar to shares of stock in a corporation. In either case, ownership usually confers the right to vote and always confers the right to share in profits.

Advantages and Disadvantages of Forming as a Limited Liability Company

The primary advantage of an LLC is that the members are not personally liable for the debts and liabilities of the LLC. The LLC allows individuals to organize with limited liability with fewer restrictions and fewer formalities that were necessary to form "S" or "C" corporations. Also, most limited liability companies can use the cash method of accounting, which means income is not generally taxed until it is received.

An LLC can be taxed either as a "pass-through" entity, like a partnership, or as a regular corporation. A regular corporation pays a corporate tax on its net income (the first tax), and then the stockholders pay income tax on dividends (the second tax) when the corporation distributes profits. With an LLC, the profits "pass through" to the owners, who pay taxes at their individual tax rates. Also, the members can deduct the business's operating losses against the member's regular income to the extent permitted by law, which can be helpful if the project anticipates losses in the first few years.

A member may become liable for LLC debts if the member personally guarantees the debts, if personal funds are intermingled with LLC funds, if the LLC has minimal insurance, or if the members do not contribute enough money to the LLC when it is formed. In order to maintain the separate form of the LLC and maintain the liability protection of its members, LLC owners must carefully maintain separate records and keep personal affairs separate from the LLCs business. In particular, the LLCs money should never be intermingled with personal money.

Formalities

Although an LLC requires fewer formalities than a corporation, there is still more paperwork involved in an LLC than a sole proprietorship or partnership. Formation paperwork (which can usually be found on the state's website) must be filed. An LLC agreement is essential to govern the relationships of the members, the financial arrangements and regulation of the transfer of membership interests, or admission of a new member. In the absence of an LLC agreement, the state's LLC laws will be applied to the LLC. In general, the name of an LLC must clearly indicate that is an LLC and end with the words "Limited Liability Company," "LLC," "L.L.C.," or "Ltd. Liability Co."

Cooperative

A cooperative is a legal entity owned and democratically controlled by its members. Members often have a close association with the enterprise as producers or consumers of its products or services, or as its employees.

A consumers' cooperative is a business owned by its customers. Employees can also generally become members. Members vote on major decisions, and elect the board of directors from a candidate pool of members.

Generally, cooperatives are organized as non-capital stock corporations under state-specific cooperative laws. However, cooperatives may also be unincorporated associations or business corporations such as limited liability companies or partnerships. Cooperatives often share earnings with the membership as dividends, which are divided among the members according to the members' participation in the enterprise (such as patronage) instead of according to the value of their shares. However, regardless of the amount of a member's contribution to the co-op, each member has only one vote. For tax purposes, most cooperatives are taxed as separate entities like corporations, though some are tax exempt.

Advantages and Disadvantages of Forming as a Cooperative

The democratic nature of cooperatives might appeal to community shared solar project organizers, based on compatible goals of creating a collaborative and accessible structure. But there are significant limitations to cooperative structures that have made them an unpopular choice for renewable energy projects. For example, the Clean Energy Collective started out as a cooperative and converted to an LLC. Traditionally, members have little input into business operations and in certain states, members have to personally benefit from the co-op's products and services (example: REI). In those states, the co-op structure is not designed to bring in outside investment from individuals that cannot partake of the co-op's products and services. However, in other states, outside investment is permitted and states are beginning to recognize the value of the co-op structure in a community shared solar setting. The costs of the documentation and filing requirements can be high.

Formalities

Usually, cooperatives are formed by filing articles of incorporation with the state. It is important to create a comprehensive set of bylaws to govern the members' relationship and the duties and obligations of the board of directors that will operate the business without significant input from the members. If the co-op is to be operated as a nonprofit entity, the co-op will need to comply with the formalities for forming such an entity.

Note Regarding the Co-op Model

While solar power production co-ops are popular in Europe, they are rare in the United States. One explanation for this discrepancy may be in the differing regulatory regimes. In the U.S., in order to reduce costs from state and federal securities compliance, co-op members receive limited compensation on capital subscribed as a condition of membership. This makes the co-op model less attractive to investors looking for a monetary return.

For-Profit Corporations

A corporation is a separate and distinct legal entity, meaning the corporation does business under its own name. A corporation issues/sells voting common stock and (sometimes) preferred stock, which can be voting or non-voting. The owners of the stock are called "stockholders" or "shareholders."

A corporation is managed by a board of directors elected by the shareholders, which is responsible for making major business decisions and overseeing the general affairs of the corporation. The directors appoint officers, who run the day-to-day operations of the corporation. Each corporation must have at least one director. In a small ("close") corporation, the shareholders, the directors and the officers are usually the same three or four people, but in a larger corporation, the shareholders are passive investors and, other than electing directors, have little control over the business operations of the corporation. In this case, the stock issued to passive shareholders can constitute a security.

Directors, officers, and in some cases, the majority shareholders of a corporation owe "duties of loyalty and care" to the corporation. Generally, this means the directors must act in good faith, with reasonable care, and in the best interest of the corporation. Directors, officers and majority shareholders must not use their positions to gain personally from transactions with corporations without complying with certain legal formalities.

Advantages and Disadvantages of Forming as a Corporation

The primary advantage of a corporation is that shareholders are not generally liable for corporate debts, provided shareholders follow their particular states' rules regarding formation of the corporate and maintenance of the corporate identity. For example, a shareholder may be liable for corporate debts if the shareholder personally guarantees the debts, if personal funds are intermingled with corporate funds, or if the corporation is undercapitalized (i.e., shareholders do not contribute enough money to the corporation when it is formed). Other actions may affect the liability of the shareholders, so anyone considering this business entity should consult a legal professional to ensure that all the proper formalities are followed. The other major disadvantage is that shares in a corporation are deemed to be securities.

A corporation can elect to be taxed either as a "C corporation" or as an "S corporation." A "C" (or regular) corporation pays a corporate tax on its net income (the first tax), and then the stockholders pay income tax on dividends (the second tax) when the corporation distributes profits. An "S" corporation is like a pass-through entity, but there are limitations on the number of shareholders and who may be a shareholder.

Formalities

A corporation is required to hold annual meetings of shareholders to elect directors. In most jurisdictions, meetings can be held in person or by electronic means that allow all persons to hear the proceedings. It is important to maintain the corporation's records scrupulously to prevent creditors making claims against the shareholders. The corporation also must obtain a separate tax identification and separate bank account.

The name of a corporation must contain words that identify the company as a limited liability entity, such as "Inc.," "Ltd.," or "Corporation."

Nonprofit Entities

A nonprofit entity can be a corporation, or other form of business entity that is organized to meet specific tax-exempt purposes. Common examples of nonprofits include: religious, charitable and political organizations, credit unions, and membership clubs such as the Elk's Club. To qualify for nonprofit status, the entity must be formed to benefit (1) the public, (2) a specific group of individuals, or (3) the membership of the nonprofit. If the nonprofit has members, it may be able to elect directors and approve a sale or merger; however, many smaller nonprofits do not have members, due to the additional paperwork and required formalities. Even without members, donors may participate as advisors, patrons, or contributors, but do not have a vote in the nonprofit's operations.

Being a nonprofit does not mean the entity cannot make a profit. Nonprofits can sell goods or services for money and can pay competitive salaries to officers and employees. The primary limitation is that any profits generated by the nonprofit's business operations cannot be distributed to members, but must be retained by the nonprofit and used to further its purposes and run its business. Nonprofits are exempt from income, sales, and property taxes and allow donors to deduct their donations from their taxes. Absent misuse of the nonprofit's resources, directors, officers, and members are not liable for the debts of the nonprofit.

Although tax-exempt entities such as nonprofits are not usually eligible for tax credits, the entities may be eligible for other grants or other sources of foundation funding that would not otherwise be available to a for profit entity.

(Note: The discussion pertains to nonprofit entities that pursue solar projects as part of their core missions. For a discussion of how an existing nonprofit may fund a solar project through donations, see Section 2, Community Shared Solar Project Models: Nonprofit Model.)

Advantages and Disadvantages of Forming as a Nonprofit Corporation

The largest advantage of organizing as a nonprofit is that the entity is exempt from paying taxes on its profits, provided the activities of the entity continue to meet the requirements for exemption. It is important to note that simply forming a nonprofit does not automatically qualify the entity for federal and state tax exemption—only an officially recognized nonprofit entity can apply for federal and state tax exemption. This application is often referred to as the 501(c)(3) application, which is the IRS code section most commonly applicable to nonprofits. In fact, there are more than 20 code sections for nonprofit qualification. Another common one is 501(c)(7), which applies to social and recreational clubs.

Formalities

Unless a nonprofit corporation files a 501(c)(3) application with the IRS, it will not be exempt from paying federal income taxes. If the nonprofit's purpose qualifies under 501(c)(3), then a legal professional can help the nonprofit prepare the application. Each state also requires a tax exempt application; however, most states accept the federal tax exempt application in place of the state's tax exemption.

The process for forming the nonprofit can take several months. Generally, the IRS takes three to five months to examine and approve the 501(c)(3) application.

Like any business entity, it is critical to maintain the separate corporate identity of the nonprofit. This entails setting up a separate bank account, maintaining good corporate records, and holding regular board meetings.

Summary Table of Business Types

Entity Type	Liability for Owners	Taxation	Primary Advantages	Primary Disadvantages
General Partnerships	Personal liability	Pass-through	Ease of formation; pass-through taxation	Personal liability of partners
Limited Partnerships	Personal liability for general partners; limited liability for limited partners	Pass-through	Pass-through taxation; limited liability for limited partners	No liability shield for general partner

Entity Type	Liability for Owners	Taxation	Primary Advantages	Primary Disadvantages
Limited Liability Companies	Limited liability	Usually pass-through	Pass-through taxation; fewer formalities to maintain the LLC structure than corporations	Relatively new structure; may be harder to get financing
Cooperatives	Limited Liability	Separate tax entity	Cooperative principles	Inflexible Structure
"S" Corporations Limited Liability	Limited Liability	Pass-through	Liability shield; ease of investment; ease of transfer of shares in larger, non-close corporations	Limitations on number and identity of members
"C" Corporation	Limited Liability	Separate tax entity	Liability shield; ease of investment; ease of transfer of shares in larger non-close corporations	Complexity; double taxation
Non- Profit Entities	Limited Liability	Separate tax identity; tax exempt	Tax-exempt; tax deduction for donors	No return for donors; business purposes are limited; no voting rights for donors

APPENDIX B

Introduction to IREC's Community Renewables Model Program Rules

Taking into account the various community shared renewables approaches that have been implemented, the Interstate Renewable Energy Council (IREC) worked closely with The Vote Solar Initiative to develop its *Community Renewables Model Program Rules*. IREC designed the Model Program Rules as a starting point to assist stakeholders in developing programs that meet the diverse needs of the communities. The Model Program Rules address issues such as renewable system size, interconnection, eligibility for participation, allocation of the benefits flowing from participation, net metering of system production, and other essential features of a community shared renewable energy program.

The first part of this process was the development of a Community Renewable Power Proposal (Proposal) to generate stakeholder input on best practices in this emerging policy area. As part of the development of the Proposal, IREC collaborated with a diverse set of stakeholders and reviewed current community shared renewables efforts taking place at the municipal and state levels, including efforts in Massachusetts, Colorado, California, Washington, and Utah.

Two key principles greatly influenced the development of the Proposal, and ultimately, the Model Program Rules.

First, as a foundational matter, IREC believes that participants in a community shared renewables program should have an experience that is as similar as possible to that of customers investing in on-site renewable energy. The reason for this is simple: on-site programs in many states have been very successful in motivating energy consumers to invest in renewable energy, so replicating the program elements that spurred this motivation is logical and builds off of foundations already in place. In particular, many customers appear to be highly motivated to zero out their monthly energy bill through net metering. In addition, customers participating in existing programs have been shown to install more energy efficiency measures than nonparticipants, because the customers are highly motivated to reduce their energy bills. On-bill net metering for community shared renewables systems can maintain a participating customer's motivation to reduce his or her energy bill and adopt energy efficiency measures.

Second, community shared renewables should be additive to successful on-site renewable energy programs and not undermine on-site renewable energy programs. Over the previous decades, renewable energy companies have invested considerable resources in building their businesses. This private investment in time and resources has helped expand markets for renewable energy in partnership with ratepayer-funded incentive programs. For this reason, it makes little sense to undermine successful on-site programs, and the businesses based upon these programs, when seeking to expand options for participation.

IREC's proposal generated significant feedback from utilities, industry participants, and other stakeholders, which was used to develop the Model Program Rules. For example, the Model Program Rules specify a renewable system size cap of 2 MW. This size cap was chosen because a 2-MW system maintains economies of scale both in the installed cost of the system and in the participation/marketing costs for a business engaged in developing community shared renewables systems, and still allows for relatively low-cost interconnection on most utility distribution systems. Regarding another program element, the minimum number of participants, IREC considered conflicting program impacts raised by stakeholders. On one hand, if a program requires too many participants, gathering up the minimum number of participants can make participation by smaller systems difficult. On the other hand, if a program requires just one participant, then the "community" aspect of a community shared renewables program is taken out of the picture. In considering these two concerns, IREC has chosen to require a minimum of two participants in a community shared renewables system. This requirement will allow duplex owners, small apartment buildings, and small commercial establishments to participate.

During discussions with stakeholders on the development of these Model Program Rules, five areas emerged as deserving of special attention:

1. Method of allocating the benefits of participation
2. Program administration
3. Financing options for community shared renewables
4. Valuation of the energy produced by the community shared renewables system
5. Utility compensation for program administration

IREC intends to continue to develop and refine its Model Program Rules. IREC anticipates issuing a revised version sometime in 2012.

Allocating the Benefits of Participation

Allocating benefits to program participants is a critical element of developing a successful renewables program. In considering the best method for allocating benefits to participating customers, IREC felt it was important to avoid structuring a program as a wholesale program that could result in taxable income. From an economic standpoint, it makes little sense for customers to invest in greening their energy supply if benefits of doing so will be siphoned off in taxes. Therefore, IREC has chosen to avoid program structures that allocate benefits in a manner that might raise these concerns, such as cutting a check for the value of energy produced and instead choosing virtual net metering (VNM) to allocate the benefits of participation onto a customer's monthly electric bill. Additionally, many customers are motivated to offset as much of their energy bills as possible. While the reasons underlying this motivation are complex, most states' existing net metering programs accommodate this desire by placing net metering credits on a customer's monthly bill. VNM maintains this direct relationship between customers' investments in renewable energy and a reduction in the customers' utility bills. Last, consistent with the principles outlined above, VNM provides a similarity in experience between customers installing on-site systems and those customers who participate in a community shared renewables program.

Program Administration

Program administration represents another critical area of program design. Existing community shared renewables programs have fallen into two categories with regard to who has program administration responsibilities: customer representatives (as in Vermont's group billing program) or utilities. IREC believes the best approach is to allow utilities to administer a community shared renewables program. This framework allows an entity with significant experience in administering complex energy programs to administer the details of a community shared renewables program, which may have many participants. Use of a utility administrator also prevents concern about creditworthiness of a third-party customer representative.

Financing Community Shared Renewables

Renewable energy systems represent significant investments. Accordingly, an array of local, state, and federal incentives have been developed to incentivize customer investment in them. To maximize the availability of funding and to ensure available incentives are used as efficiently as possible, IREC's Model Program Rules support direct ownership, third-party ownership, and utility ownership of community shared renewable systems. Allowing a multitude of ownership options will maximize the availability of funding and ensure federal, state and local incentives are used to the fullest extent. Of particular note, third-party ownership of a renewable energy system can be critical to tapping into funders who are able to use all available federal tax credits. The efficient use of federal tax credits can result in a reduction in the cost of renewable energy by almost 50 percent. Recognizing the important role third-party ownership can play in increasing access to renewable energy, thirteen states have explicitly authorized third-party ownership of on-site renewable energy systems. In addition, legislation enacting community shared renewables programs in Colorado, Massachusetts, Delaware and Washington have made clear that third-party owners of community shared renewable energy systems are not subject to public utility regulation.

An important aspect of allowing utility ownership is a requirement that all system purchase costs, operation and maintenance costs, necessary investment returns, and other costs related to a utility-owned system must be recovered from participants enrolled in a utility program. This requirement is important to maintaining a level playing field between utility offerings and offerings of other parties. It ensures that all costs incurred by a utility to operate a community shared renewable system are recovered from program participants (as in the case with other competitive providers) and not from non-participating ratepayers.

Valuation of the Energy Produced by the Renewable System

At the heart of a successful community shared renewables program is the experience participants have as a result of their participation in a project. With industry input, the regulators must make a threshold decision on whether the net metering credits generated by a project should be transferred to participants as a 1:1 kWh offset on the customer's utility bill or whether the kilowatt-hours should be given a monetary value based on some other rate. This is important because it determines whether the value of a credit can be administratively determined or whether the value will be different for each participant and be based on the amount that a participant would otherwise pay per kWh of electricity provided by a utility.

Under most states' net metering programs, net metering credits generated by an on-site system are used to directly offset killowatt-hours delivered by a utility when a customer-generator's consumption exceeds the energy supplied by a renewable energy system. Given that most customer-generation is simply used on site without requiring that a customer's billing meter spin backwards to earn net metering credits, this framework makes intuitive sense. However, the vast majority of participants in community shared solar projects will not have generation located behind a billing meter, so the link between excess production and 1:1 kWh offsets is not as important. In addition, it can be more difficult to administer this arrangement once a generation source is separated from the participants who would like to receive electricity from that system. Providing kWh credits can be particularly difficult to track if a customer is on a time-of-use rate structure because kWh production would have to be tracked within time periods and applied to the customers' bills within time periods.

Credits denominated in dollars and cents are often much easier for utilities to administer and often require fewer billing software changes. Accordingly, for ease of administration by utilities, IREC chose to allow kWh generated by a project to be given a monetary value that can be applied to participants' bills. In determining the appropriate monetary value to assign to kWh credits, three approaches are currently in use for community shared solar projects: (1) valuing a kWh credit based on the retail rate in effect where the project is located (MA does this); (2) valuing a credit based on a the retail rate in effect for the participant (CA does this); or (3) valuing a credit based on some other approach, such as the wholesale value of power production (Maine's approach).

IREC chose the second approach for several reasons. First, valuing the kWh credit at the retail rate in effect for the participant maintains the ability of the project to act as a price hedge against future utility rate increases. Second, valuing the kWh credit at the participant's retail rate maintains an outcome that is as close as possible to the experience participants would have if they installed a solar energy system on-site. Third, transforming the kWh credit into a monetary credit should simplify the calculations required for customers that need to compensate a utility for the use of the distribution system. Finally, transforming kWh credits into a monetary credit allows customers that face demand charges to have their participation in solar generation recognized by valuing their kWh credits at a total aggregate retail rate.

Compensating Utilities for Program Administration

One of the most complex issues with development of community shared renewables programs is setting an appropriate compensation rate for utilities to administer programs. It should be relatively noncontroversial that utilities should be allowed to recoup administrative fees. However, the propriety of allowing a utility to recover costs for distribution service is a more controversial topic, and one on which California and Massachusetts have taken different approaches.

In Massachusetts, net metering credits created by a "neighborhood net-metered facility" do not contain the distribution portion of a fully bundled retail rate. As a result, participants in a "neighborhood" facility continue to pay distribution charges. However, participants do not pay transmission fees. Currently, the Massachusetts approach seems reasonable because neighborhood net-metered facilities are limited to 2 MW and participating customers may be located anywhere within a distribution utility's service territory. Although participating systems will be located close to load with no use of the transmission system, a utility would only need to be compensated for use of the distribution system.

Unlike Massachusetts, in California, net metering credits are valued at a fully bundled retail rate. This outcome appears sensible because, unlike the Massachusetts' program, California's virtual net metering program is available only to occupants of certain types of multi-tenant buildings. Thus, California participants will be located within the same building on the same distribution circuit and, as a consequence, use of the distribution system will be nonexistent or minimal.

IREC's Model Program Rules take a nuanced approach to this issue by specifying that customers on the same distribution circuit as the community shared solar project will have their killowatt-hour credits valued at the full retail rate, while also allowing a stakeholder process to determine an appropriate level of compensation for use of a utility's distribution system once a number of factors have been taken into account.

IREC's Community Shared Renewables Model Program Rules

These rules were created by the Interstate Renewable Energy Council and The Vote Solar Initiative to serve as a guide for renewable energy stakeholders to consider when developing community shared renewables policies to meet the needs of their states. The rules provide a framework for building a community shared renewables program that is additive to successful on-site renewable energy programs and uses solar, wind, hydro, biomass and other renewable energy sources to allow communities to promote local job growth. These program rules are solely the recommendation of the Interstate Renewable Energy Council and The Vote Solar Initiative and do not necessarily reflect the recommendation of the authors, DOE, or NREL.

I. Definitions. As used within these rules, unless the context otherwise requires

(a) "Biomass" means a power source that is comprised of, but not limited to, combustible residues or gases from forest products manufacturing; waste, byproducts, or products from agricultural and orchard crops; waste or co products from livestock and poultry operations; waste or byproducts from food processing, urban wood waste, municipal liquid waste treatment operations, and landfill gas.[22]

(b) "Community Energy Generating Facility" means Renewable Energy Generation that is interconnected at the distribution system level and that is located in or near a community served by an Electricity Provider where the electricity generated by the facility is credited to the Subscribers to the facility. A Community Energy Generating Facility may be located either as a stand-alone facility, called herein a stand-alone Community Energy Generating Facility, or behind the meter of a participating Subscriber, called herein a hosted Community Energy Generating Facility. A Community Energy Generating Facility may be no larger than two megawatts (MW). A Community Energy Generating Facility must have at least two Subscribers.

(c) "Electricity Provider" means the jurisdictional entity that is required to offer Net Metering service to Subscribers pursuant to [code section for applicable Net Metering rules].

(d) "Locational Benefits" mean the benefits accruing to the Electricity Provider due to the location of the Community Energy Generating Facility on the distribution grid. Locational Benefits include such benefits as avoided transmission and distribution system upgrades, reduced transmission and distribution level line losses, and ancillary services.

(e) "Net Metering" means a methodology under which electric energy generated by or on behalf of a Subscriber and delivered to the Electricity Provider's local distribution facilities may be used to offset electric energy provided by the Electricity Provider to the Subscriber during the applicable billing period.

(f) "Renewable Energy Credit" means a tradable instrument that includes all renewable and environmental attributes associated with the production of electricity from a Community Energy Generating Facility.

(g) "Renewable Energy Generation" means an electrical energy generation system that uses one or more of the following fuels or energy sources: Biomass, solar energy, geothermal energy, wind energy, ocean energy, hydroelectric power, or hydrogen produced from any of these resources.

(h) "Subscriber" means a retail customer of a utility who owns a Subscription and who has identified one or more individual meters or accounts to which the Subscription shall be attributed. Such individual meters or accounts shall be within the same Electricity Provider's distribution service territory as the Community Energy Generating Facility.

(i) "Subscriber Organization" means an organization whose sole purpose is to beneficially own and operate a Community Energy Generating Facility for the Subscribers of the Community Energy Generating Facility. A Subscriber Organization may be any for-profit or non-profit entity permitted by [state] law. The Community Energy Generating Facility may also be built, owned, and operated by a third party under contract with the Subscriber Organization.

(j) "Subscription" means an interest in a Community Energy Generating Facility. Each Subscription shall be sized to represent at least one kilowatt of the Community Energy Generating Facility's generating capacity; provided, however, that the Subscription is sized to produce no more than 120% of the Subscriber's average annual electrical consumption. For Subscribers participating in meter aggregation, 120% of the Subscriber's aggregate electrical consumption may be based on the individual meters or accounts that the Subscriber wishes to aggregate pursuant to these rules. In sizing the Subscription, a deduction for the amount of any existing renewable energy generation at the Subscriber's premises or any Subscriptions owned by the Subscriber in other Community Energy Generating Facilities shall be made.

(k) "Total Aggregate Retail Rate" means the total retail rate that would be charged to a Subscriber if all electric rate components of the Subscriber's electric bill, including any riders or other additional tariffs, except for minimum monthly charges, such as meter reading fees or customer charges, were expressed as per kilowatt-hour (kWh) charges.

II. General Provisions

(a) Subscriptions in a Community Energy Generating Facility may be transferred or assigned to a Subscriber Organization or to any person or entity that qualifies to be a Subscriber under these rules.

(b) New Subscribers may be added at the beginning of each billing cycle. The owner of a Community Energy Generating Facility or its designated agent shall inform the Electricity Provider of the following information concerning the Subscribers to the Community Energy Generating Facility on no more than a monthly basis: (1) a list of individual Subscribers by name, address, account number; (2) the proportional interest of each Subscriber in the Community Energy

Generating Facility; and (3) for Subscribers who participate in meter aggregation, the rank order for the additional meters or accounts to which Net Metering credits are to be applied.

(c) A Subscriber may change the individual meters or accounts to which the Community Energy Generating Facility's electricity generation shall be attributed for that Subscriber no more than once quarterly, so long as the individual meters or accounts are eligible to participate.

(d) An Electricity Provider may require that customers participating in a Community Energy Generating Facility have their meters read on the same billing cycle.

(e) If the full electrical output of a stand-alone Community Energy Generating Facility or the excess generation from a hosted Community Energy Generating Facility is not fully allocated to Subscribers, the Electricity Provider shall purchase the unsubscribed energy at a kWh rate that reflects the full value of the generation. Such rate shall include the avoided cost of the energy, including any Locational Benefits of the Community Energy Generating Facility.

(f) If a Subscriber ceases to be a customer within the distribution service territory within which the Community Energy Generating Facility is located, the Subscriber must transfer or assign their Subscription back to their Subscriber Organization or to any person or entity that qualifies to be a Subscriber under these rules.

(g) If the Subscriber ceases to be a customer of the Electricity Provider or switches Electricity Providers, the Electricity Provider is not required to provide compensation to the Subscriber for any unused Net Metering credits.

(h) A Community Energy Generating Facility shall be deemed to be located on the premises of each Subscriber for the purpose of determining eligibility for state incentives.

(i) Neither the owners of, nor the Subscribers to, a Community Energy Generating Facility shall be considered public utilities subject to regulation by the [responsible agency having regulatory oversight] solely as a result of their interest in the Community Energy Generating Facility.

(j) Prices paid for Subscriptions in a Community Energy Generating Facility shall not be subject to regulation by the [responsible agency having regulatory oversight].

(k) A Subscriber owns the Renewable Energy Credits (RECs) associated with the electricity allocated to the Subscriber's Subscription, unless such RECs were explicitly contracted for through a separate transaction independent of any Net Metering or interconnection tariff or contract. For a Community Energy Generating Facility located behind the meter of a participating Subscriber, the host Subscriber owns the RECs associated with the electricity consumed on-site, unless the RECs were explicitly contracted for through a separate transaction independent of any Net Metering or interconnection tariff or contract.

(l) The dispute resolution procedures available to parties in the Electricity Provider's interconnection tariff shall be available for the purposes of resolving disputes between an Electricity Provider and Subscribers or their designated representative for disputes involving the Electricity Provider's allocation of Net Metering credits to the Subscriber's electricity bill consistent with the allocations provided pursuant to Rule II.b. The Electricity Provider shall not be responsible for resolving disputes related to the agreements between a Subscriber, the owner of a Community Energy Generating Facility, and/or a Subscription Organization or any other party. This provision shall in no way limit any other rights the Subscriber may have related to an Electricity Provider's provision of electric service or other matters as provided by, but not limited to, tariff, decision of [responsible regulatory body or agency], or statute.

III. Net-Metering Provisions

(a) An Electricity Provider shall not limit the cumulative, aggregate generating capacity of Community Energy Generating Facilities.[23]

(b) For a Community Energy Generating Facility, the total amount of electricity expressed in kWh available for allocation to Subscribers, and the total amount of RECs generated by the Community Energy Generating Facility and allocated to Subscribers, shall be determined by a production meter installed and paid for by the owner(s) of the Community Energy Generating Facility. It shall be the Electricity Provider's responsibility to read the production meter.

(c) For a hosted Community Energy Generating Facility, the determination of the quantity of kWh credits available to Subscribers of that facility for Net Metering, including the host Subscriber, shall be based on any energy production of the Community Energy Generating Facility that exceeds the host Subscriber's instantaneous on-site consumption during the applicable billing period and the Subscribers' Subscriptions in that Community Energy Generating Facility.

(d) For a stand-alone Community Energy Generating Facility, the determination of the quantity of kWh credits available to each Subscriber of that Community Energy Generating Facility for Net Metering shall be based on the total exported generation of the Community Energy Generating Facility and each Subscriber's Subscription in that Community Energy Generating Facility.

(e) For Subscribers that host a Community Energy Generating Facility or where participating Subscribers are located on the same distribution feeder as the Community Energy Generating Facility, the value of the kWh credits for the host Subscriber and those Subscribers on the same distribution feeder shall be calculated by multiplying the Subscriber's share of the kWh electricity production from the Community Energy Generating Facility by the retail rate for the Subscriber. For Subscribers on tariffs that contain demand charges, the retail rate for the Subscriber shall be calculated as the Total Aggregate Retail Rate for the Subscriber.

(f) For all other Subscribers to a Community Energy Generating Facility, value of the kWh credits allocated to each Subscriber shall be calculated by multiplying the Subscriber's share of the electricity production from the Community Energy Generating Facility by the retail rate as charged to the Subscriber, minus a reasonable charge as determined by the [responsible agency having regulatory oversight] to cover the Electricity Provider's costs of delivering the electricity generated by the community electricity generating facility to the Subscriber's premises after taking into account the Locational Benefits and other benefits2 provided by the Community Energy Generating Facility. The [responsible agency having regulatory oversight] shall ensure that this charge does not reflect costs that are already recovered by the Electricity Provider from the Subscriber through other charges. In no event, shall the charge, if assessed, be greater than the Subscriber's distribution service charge as determined on a per kWh basis.

(g) The Electricity Provider shall carry over any excess kWh credits earned by a Subscriber and not used in the current billing period to offset the Subscriber's consumption in subsequent billing periods until all credits are used. Any excess kWh credits shall not reduce any fixed monthly customer charges imposed by the Electricity Provider.

End Notes

[1] Supply Curves for Rooftop Solar PV-Generated Electricity for the United States, National Renewable Energy Laboratory, Nov. 2008. www.nrel.gov/docs/fy09osti/44073.pdf .

[2] Growing a Green Economy for All: From Green Jobs to Green Ownership, The Democracy Collaborative, June 2010, p. 22. www.community-wealth.org/_pdfs/news/recent-articles/07-10/report-warren-dubb.pdf.

[3] ITC tax benefits may not be readily accessible to for-profit utilities, due to the normalization accounting rules.

[4] P. Alvarez and B. Hodges. (2009). "Buying Into Solar." Public Utilities Fortnightly. p. 57.

[5] 26 USC 136 states that subsidies from public utilities for energy conservation measures are not taxable. For example, Washington State's production incentive was ruled to be not income. See http://apps.leg.wa.gov/WAC/default.aspx?dispo=true&cite=458-20.

[6] SB1 is the California Solar Initiative, a state mandate requiring all California electric utilities to offer a 10-year program of declining incentives for customer-sited PV. It expires at the end of 2016.

[7] For a list of IRS material participation tests and other details about passive activity and at-risk rules, see IRS Publication 925, available at: www.irs.gov/pub/irs-pdf/p925.pdf .

[8] To be considered an accredited investor, an individual must have either: 1) a net worth of more than $1 million or 2) an annual income of $200,000 ($300,000 jointly with a spouse) in each of the most recent two years and a reasonable expectation of having the same income level in the current year.

[9] Tangerine Power, LLC, based in Washington State has created a business model for a solar power co-op and has launched the Edmonds Community Solar Cooperative.

[10] The Portland Habilitation Center Northwest, a nonprofit organization, partnered with U.S. Bancorp Community Development Corporation, which will own and finance an 870 kW system to provide energy to the nonprofit.

[11] See Vermont Public Service Board Rule 5.100, available at: www.psb.vermont.gov/sites/psb/files/rules/OfficialAdoptedRules/5100adoptedrule_2.pdf.

[12] Vermont Energy Act of 2011, H. 56, 30 V.S.A. 219a(a)(3)(A), available at: www.leg.state.vt.us/docs/2012/bills/Passed/H-056.pdf.

[13] See Colorado House Bill 10-1342, available at: www.leg.state.co.us/ .

[14] 4 C.C.R. 723-3 Rule 3664, available at: www.dora.state.co.us/puc/rules/723-3.pdf .

[15] S.B. 267, An Act to Amend Title 26 of the Delaware Code Relating to Net Energy Metering, July, 2010. www.legis.delaware.gov/LIS/LIS145.NSF/db0bad0e2af0bf31852568a5005f0f58/f17ba623105f222b85257745007 65d6e?OpenDocument ; DE PSC, Order No. 7984, June, 2011. www.depsc.delaware.gov/orders/7984.pdf.

[16] See Multifamily Affordable Solar Housing Semiannual Report, available at: www.cpuc.ca.gov/NR/rdonlyres/B3644285-F573-428F-AA0A-A2497A30401B/0/MASHSemiAnnualReport.pdf.

[17] See California Solar Initiative Low-Income Solar Program Evaluation, available at: www.cpuc.ca.gov/NR/rdonlyres/13AAEDF8-BB7D-4FBD-AC05-3FC2B9CBF746/0/CSISASH_MASHImpact_and_Cost_Benefit_Report.pdf .

[18] See California Solar Initiative Phase One Modifications, Decision 11-07-031, Rulemaking 10-05-004, available at: www.docs.cpuc.ca.gov/word_pdf/FINAL_DECISION/139683.pdf .

[19] See An Act To Establish the Community-based Renewable Energy Pilot Program, available at: www.mainelegislature.org/legis/bills/bills_124th/chapters/PUBLIC329.asp .

[20] Financing Non-Residential Photovoltaic Projects: Options and Implications, Lawrence Berkeley National Laboratory, Jan 2009. http://eetd.lbl.gov/ea/emp/reports/lbnl-1410e.pdf.

[21] See Utah Code 59-7-614.3, available at: http://www.le.state.ut.us/~code/TITLE59/htm/59_07_061403.htm .

[22] The definition of Biomass may need to be adjusted to reflect state renewable portfolio standard definitions.

[23] This program rule is based upon IREC's Net Metering Model Rule (b)(2), which specifies that the cumulative, aggregate generating capacity net metered by on-site renewable generation facilities shall not be arbitrarily limited. Some states cap the total amount of aggregate Renewable Energy Generation that can be Net Metered for a particular Electricity Provider. Most commonly, aggregate enrollment caps are expressed as a percentage of an Electricity Provider's peak demand based on the aggregate of nameplate capacity of the generation systems (though it should be noted that capacity calculations are not standardized in their methodology across or even within states). Such percentages can vary from as low as 0.1% to as high as 20%. IREC believes aggregate caps arbitrarily and unnecessarily limit private investment in Renewable Energy Generation and needlessly curtail the flow of benefits that are associated with customer-side Renewable Energy Generation. For states that place an aggregate enrollment cap on net metered generation, that cap should be removed or expanded to ensure that community renewables programs do not undermine successful on-site programs.

In: Shared Solar Power
Editors: E. J. Rose and D. Marks
ISBN 978-1-62257-527-5

Chapter 2

THE SOLARIZE GUIDEBOOK: A COMMUNITY GUIDE TO COLLECTIVE PURCHASING OF RESIDENTIAL PV SYSTEMS*

U.S. Department of Energy

ACKNOWLEDGMENTS

This guide is an updated version of the original The Solarize Guidebook, published in February 2011 (see www.nrel.gov/docs/fy11osti/50440.pdf), which was developed for the National Renewable Energy Laboratory and the City of Portland. The original Solarize campaigns were initiated and replicated by Portland's Neighborhood Coalition network with help from the Energy Trust of Oregon, City of Portland, and Solar Oregon.

INTRODUCTION

The first "Solarize" campaign started as a grassroots effort to help residents of Portland, Oregon, overcome the financial and logistical barriers

* This is an edited, reformatted and augmented version of the U.S. Department of Energy, DOE/GO-102012-3578, dated May 2012.

to installing solar power. What began in one neighborhood as "Solarize Southeast!" quickly caught on with residents across the city. With support from the U.S. Department of Energy's (DOE) Solar America Communities program, the City of Portland Bureau of Planning and Sustainability partnered with neighborhood coalition offices, Solar Oregon (the state American Solar Energy Society chapter) and Energy Trust of Oregon to provide community organizing, technical assistance, project management, and rebates in a wildly popular grassroots-driven program. After three years of Solarize campaigns, Portland has added over 1.7 MW of distributed photovoltaics (PV) and established a strong, steady solar installation economy.

Since the publication of the first Solarize Guidebook in 2011, dozens of communities, companies and contractors across the U.S. have launched their own versions of a neighborhood collective purchasing program. With installed costs for behind-the-meter (distributed) solar dropping 17% in 2010 and continuing to fall in 2011,[1] the residential PV market in the U.S. is poised to continue expansion and Solarize campaigns can accelerate this growth.

Purpose

This guidebook is intended to be a roadmap for project planners and solar advocates who want to create their own successful Solarize campaigns. It describes the key elements of the Solarize Portland campaigns and variations from projects across the country, along with lessons learned and planning templates.

The guidebook is funded by the DOE SunShot Initiative, a collaborative national initiative to make solar energy cost competitive with other forms of energy by the end of the decade.

Background and Overview of Model

The First Solarize Campaign

The first Solarize campaign began with local Portland residents who wanted to install solar power, but didn't know where to start. They imagined that if they could organize a group of neighbors to "go solar" together, they could collectively make an informed purchase and negotiate a volume discount. They turned to the local neighborhood coalition, Southeast Uplift, for assistance. Southeast Uplift approached Energy Trust of Oregon for technical and program planning support. By coincidence, Energy Trust had developed a solar PV volume purchasing program and was eager to test the model. With community volunteers, neighborhood association staff, and Energy Trust support and rebates, the first Solarize campaign was born.

Within six months of starting their campaign, Solarize Southeast had signed up more than 300 residents and installed solar on 130 homes. The 130 installations added 350 kilowatts of new PV capacity to Portland and created 18 professional-wage jobs for site assessors, engineers, project managers, journeyman electricians, and roofers.[2]

The neighborhood collective purchase concept spread quickly. With support from a DOE Solar America Communities grant,[3] the City of Portland's Bureau of Planning and Sustainability helped other community organizations take Solarize Portland to their neighborhoods, eventually completing projects that encompassed every neighborhood in the City. Taken together, these follow-on projects produced another 400 Solarize installations in 2010, increasing total PV installations almost 400% over the previous year. In 2011, the number of Solarize campaigns and installations fell, but overall, PV installations in Portland remained high, with an increasing number of non-Solarize installations. The recent introduction of solar lease and Power Purchase Agreement (PPA) options has proven very attractive. As the Solarize campaigns wrap up, these financing options are creating new demand, building a strong, steady solar market in the Portland metropolitan area. (Examples of these third-party ownership models are discussed in General Lessons and Considerations: Financing.) Although Solarize installations trended down in 2011 due to fewer campaigns and the introduction of the solar lease and PPA, independent installations have ramped up considerably following the market kick-start provided by Solarize.

Overcoming Market Barriers

Although the volunteer organizers of the first campaign did not set out to transform the market, their program design resulted in spectacular market growth. The Solarize model tackles three major market barriers: **cost, complexity**, and **customer inertia**.

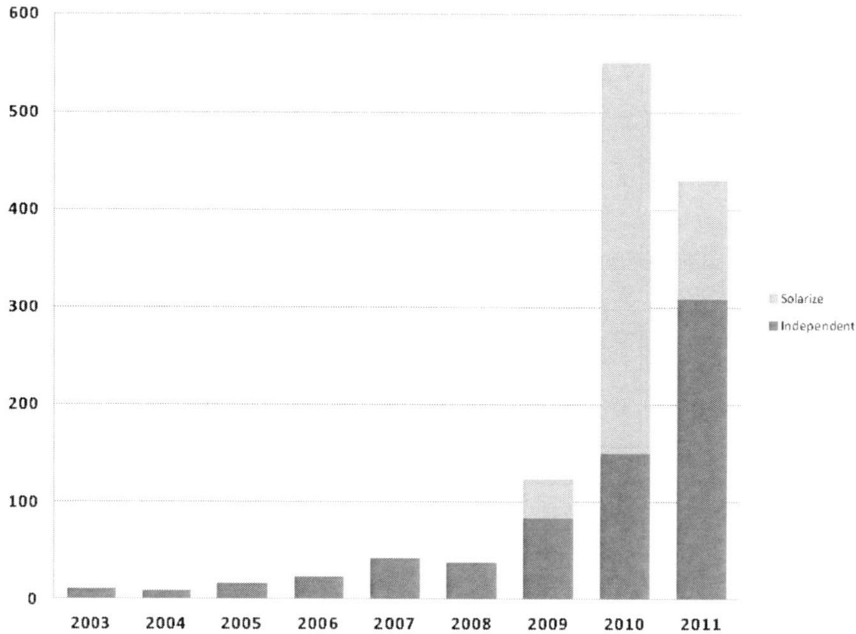

Annual Portland Residential PV Installations.

High Upfront Cost

Residential solar installations have high upfront costs. Before the first Solarize campaign launched, the upfront cost for a 3-kW system in the Portland market was approximately $27,000. By presenting a full package of federal and state tax credits and utility cash incentives, the Solarize campaign showed that the final costs were much lower than the initial sticker price. Contractor savings on marketing and lead generation drove costs down by an additional 30 to 35%.[4] A typical 3 kW installation in the first Solarize project cost only about $2,000 after tax credits and incentives.

Complexity

For many, a solar purchase seems a daunting and complex decision, involving choices about technical issues such as inverter efficiency, PV modules, and optimal array tilt. Even choosing between contractors can be an overwhelming task for those not technically inclined. Every aspect of the Solarize program was designed to provide actionable information while reducing complexity. A committee of neighbors pre-selected the contractor through a competitive bidding process and negotiated the cost. Workshops and

Q&A sessions focused on the practical steps to making a purchase. The program reduced a dizzying array of technical choices to one simple question for participants: yes or no?

Customer Inertia

The sales cycle for solar is usually more than two years from first inquiry to installation. The Solarize project overcame customer inertia to get installations in three to six months. By presenting a highly competitive price in a limited-time offering, the campaign motivated customers to act. In addition, the spirit of group endeavor afforded safety in numbers, so that participants didn't feel that they were making the decision on their own.

Essential Elements of the Solarize Model

In Portland, each successive Solarize campaign looked slightly different, reflecting the different priorities and goals of each neighborhood, but there were some common elements that led to success: competitive contractor selection, community-led outreach with a trusted community partner, and a limited-time offering.

Competitive Contractor Selection

Selecting the contractor(s) through a competitive process led by community volunteers is essential on several fronts. First, it affords homeowners the simplicity of a pre-selected contractor while building confidence that the contractor was selected from a range of options. Second, it provides a transparent process that builds customer and contractor trust. Although the criteria for selection may vary from campaign to campaign, they should reflect the particular values of the community, whether they are creating local jobs or driving prices down. By having a competitive process with clear criteria, the project organizers can justify their choice while sending a clear market message about customer and community values.

Community-Led Outreach and Education

Another element of a successful campaign is community-led outreach supported by a trusted local organization. In Portland, neighbors distributed fliers, built and updated the program website, and spoke at workshops, delivering a direct appeal from one friend to another to join the campaign. The volunteers were supported by a Neighborhood Coalition, which had a long history of helping people and a high level of community trust. Harnessing community power in this way has many benefits: The community becomes invested in the success of the project, the scope and scale of the outreach is amplified, and neighbors are more responsive to the appeals. Community-led outreach also allows the contractor to save on marketing costs because the company does not need to spend as much time generating leads. With neighborhood volunteers generating hot leads, the contractor can focus on site assessments and installations.

"Thanks to the community outreach, we saved 30% on marketing!"

Rob LaVigne, Solar City

Limited-Time Offer

Nothing motivates people like a deadline. A Solarize campaign is a limited-time offer, creating a sense of urgency among residents who don't want to miss a good deal. The limited-time offer also keeps the program true to its market transformation goals: to jump-start the solar market and then step aside. Some contractors may object to the perceived "monopoly" awarded to those selected for the project. The limited-time offer may help mitigate this contractor concern. In fact, a successful Solarize campaign can also increase business for non-Solarize installers. Installation numbers from Energy Trust demonstrate that Portland actually experienced an increase in non-Solarize installations during the Solarize campaigns.

This chart shows the 2011 pricing for a typical Solarize Portland project

3 kW PV System	Cost	Notes
Total System Cost Before Incentives	$18,000	$6.00/watt[5]
Energy Trust of Oregon Cash Incentive	($5,250)	$1.75/watt
Out of Pocket	$12,750	
Federal Tax Credit – 30%	($3,825)	Calculated after Energy Trust incentive
Oregon Residential Tax Credit	($6,000)	$2.10/DC watt; taken over 4 years
Final Cost After 4 Years	**$2,925**	

Community Feedback Spurs Innovation and Improvement

The first neighborhood collective purchasing effort in Southeast Portland was an unprecedented success, resulting in 130 new residential PV systems in six months. Although the community response was overwhelmingly positive, there were some lessons learned. A formal program evaluation commissioned by Energy Trust showed that project organizers were unprepared for the volume of customer interest. The neighborhood coalition office struggled to process customer information manually. In addition, organizers held enrollee information until the end of the enrollment period, and then gave the leads to the contractor in one batch. While this allowed the contractor to know the final price (which depended on the volume of sign-ups) before contacting the customers, it meant that the contractor received all 300 sign-ups at once. Customer follow-up time suffered, and the contractor faced boom and bust cycles. The number one suggestion for future programs, expressed by 42% of respondents, was that contractor follow-up could be faster. Project organizers took several steps to improve contractor response times, including:

- **Developing an electronic process to automate data collection and reporting.** This allowed the project organizers and the contractors to quickly see the status of every enrollee, track follow-up time, and improve customer service.
- **Processing leads on a rolling basis.** Rather than wait for the final enrollment numbers, organizers began site assessments, sales, and installations for subsequent projects before the end of enrollment. Not only did this build enthusiasm for the project by showing immediate results, but it also helped the contractor spread the work over a longer period, providing stability.

"Our mission is community- building. The Solarize project allowed people to get their hands on something and work together to make great things happen."

Tim O'Neal, Sustainability Coordinator, SE Uplift

SOLARIZE 1.0

The Basic Program

A basic neighborhood collective purchase program is designed to lead the customer through a simple process, from awareness to installation, over the course of six months. The process includes:

- **Awareness**: The grassroots campaign is advertised in fliers, emails, neighborhood newsletters, blogs, local events, and by word of mouth. Earned media from TV, radio, and newspapers can also boost awareness.
- **Education**: Workshops and Q&A sessions are offered throughout the community to allow all interested neighbors a chance to ask questions in a supportive environment and to detail the steps for participation.
- **Enrollment**: Residents are enrolled in the program through an online registration page. A short questionnaire at this time can help enrollees self-screen for solar suitability.
- **Site Assessment**: The installation contractor provides site assessments and bids to all enrollees.
- **Decision**: The customer decides whether to accept the contractor's bid at the Solarize program price. If using descending price tiers, the contractor may ask the customer to accept the current price tier, with the guarantee of a discount on their final invoice if volume targets are met. The intent is to offer few variables, so the customer's decision can be a simple "yes" or "no."
- **Installation**: The contractor installs the system and helps the customer through the paperwork for the cash incentives and any state and federal tax credits.

The Partners

A successful Solarize campaign requires the coordinated efforts of many community players. Sample campaign partners may include:

- **Trusted Nonprofit:** A community-based organization with an established history can provide credibility and institutional support.

For example, in Portland, staff members from the Neighborhood Coalition offices devoted a portion of their time to manage the campaigns. They played a crucial role in managing volunteers and reaching out to involve other supporting partners.

- **Technical Advisor:** Every campaign needs a technical advisor to help evaluate the potential solar contractors and ensure quality control. In Portland, in addition to providing the template for program design, Energy Trust created a Request for Proposals (RFP) and presented the technical tax credit and financing workshops. On the back end, Energy Trust verified that each installation met its solar requirements before issuing cash incentives.
- **Project Organizer:** An organization with institutional access can serve as the project organizer, helping to coordinate the outreach and education, contractor follow-up, and overall project timeline. In Portland, the City played this role. In addition, the City provided technical support on the RFP and created a streamlined online process for solar permitting, with a one-day turnaround on prescriptive path systems, while the City's Bureau of Development Services inspected all systems.
- **Solar Industry Organization:** A local chapter of the American Solar Energy Society (ASES), the Solar Energy Industries Association (SEIA), or a nonprofit solar industry organization can be a valuable partner. In Portland, the ASES chapter, Solar Oregon, created a database for capturing enrollees and monitored customer progress. In addition, they provided staff and volunteer Solar Ambassadors to present and offer testimonials at workshops. Other campaigns have coordinated with a local SEIA chapter to ensure that contractors know of the opportunity and have a place to discuss issues of market fairness.

Case Studies

Every successful Solarize effort is tailored to the unique features of the local market and reflects its particular community values. Indeed, allowing for this expression of values is what makes the Solarize model so attractive and empowering for participants. The following case studies offer an overview of local variations and lessons learned. These include prominent examples of everything from grassroots community-led campaigns to innovative commercially-led campaigns.

Solarize Portland: Building a Local Solar Economy

Campaigns: Six campaigns, in Southeast (2), Northeast, Southwest, North, Northwest

Installations: 560 homes

Total Installed Capacity: 1.7 MW

The Solarize Portland campaigns of 2009 through 2011 revolutionized the market for solar, driving down market prices by more than 30% across the board and generating over 50 permanent green jobs for site assessors, engineers, project managers, journeyman electricians and roofers. In 2011, the market was revolutionized once more with the entry of the solar lease and prepaid PPA options, which have subsequently been used in well over half of new PV installations. As the City looks beyond 2011, the local Solarize campaigns are winding down, but independent solar installations are increasing and the City continues to support replication efforts across the State of Oregon.

The Portland campaigns supported enduring and sustainable market growth in several ways:

1. Encouraging local contractors to respond to the RFP through a partnership or awarding the contract to two companies to split the jobs.
2. Limiting the duration of campaigns, so that the market would not be tapped out, but rather, primed for further activity.
3. Providing technical assistance to ensure that the selected contractors provided maximum benefit to the local community through local hiring, sourcing, and job training.

In addition, the citizen volunteers who were empowered to select solar installers on behalf of their neighbors chose to "share the wealth": the Portland campaigns ultimately tapped three different solar contractors and numerous local subcontractors.

Lessons and Considerations:

- *Contractor Memorandum of Understanding (MOU) promotes local hiring:* Northeast Coalition of Neighborhoods signed an MOU requiring the installation contractor to coordinate with three community-based pre-apprenticeship programs that train people to enter the construction trades. The contractor ultimately hired eight of the 18 hires from these programs in the Northeast neighborhood.
- *Using small contractors*: Small contractors need support to develop customer service mechanisms such as a customer tracking database. In addition, contractors are independent businesses and do not generally partner with one another. Rather than ask for collaboration, the RFP committee might decide to award half the jobs to one contractor and half to another, as they did in Salem, Oregon.
- *Using local manufacturers*: Project organizers in Southwest Portland wanted to "buy local."[6] The contractor suggested using panels and inverters made in Oregon. While customers were offered an option to purchase out-of-state components (because the locally manufactured products were more expensive), almost all chose the locally manufactured products, magnifying the economic impact of the program in Oregon.
- *Rising solar awareness fuels broad growth*: Although some contractors were concerned that they would suffer loss of market share when they were not selected as "Solarize" installers, the data from Portland show that non-Solarize installations also increased significantly during the Solarize campaigns. Not only did Solarize build awareness of solar energy as an option, but it also stoked demand and provided a publicly respected benchmark price, accelerating customer decision-making across the board.

For more information: Lee Rahr, City of Portland, lee.rahr@portlandoregon.gov

Solarize Washington: Simplifying Complex Pricing

Campaigns: 2 complete (Queen Anne, Magnolia), 2 in progress (Northeast Seattle, Stanwood/Camano)

Installations: 56
Total Installed Capacity: 237 kW

Seattle-based nonprofit Northwest Sustainable Energy for Economic Development (Northwest SEED) launched a Solarize program for Washington residents in early 2011. Northwest SEED works with membership-based community groups to organize Solarize campaigns neighborhood-by-neighborhood. The organization has completed two successful campaigns and began two more in 2012. Due to the structure of the Washington State Renewable Energy Production Incentive, solar systems manufactured in-state achieve a much quicker payback than solar systems manufactured out of state, despite their higher upfront cost. Selected installers offer both options to Solarize participants, adding complexity to the choice, but numbers indicate that contractors are framing the choice to match their preference at the point of sale. In addition, after the success of the first campaign, contractors are confident enough to offer a low flat price from the first sale, dispensing with participation-based pricing tiers.

As demand for Solarize campaigns has grown, Northwest SEED has begun to issue a Call for Partners to competitively select neighborhoods as hosts for upcoming Solarize campaigns. In addition, several local utilities have seen the success of Solarize Washington campaigns and are now offering support to expand the program.

Lessons and Considerations:

- *Nonprofit-led campaigns:* As a nonprofit organization, Northwest SEED faced funding challenges in moving Solarize Washington forward. Initially the organization relied heavily on foundation grants to support the program. As grant funding wanes, Northwest SEED is partnering with local utilities to provide ongoing support to Solarize campaigns. Learning from other Solarize efforts around the nation, Northwest SEED may incorporate a lead-generation fee to help create a more predictable, stable funding source for future campaigns.
- *Contractor-led projects:* Following the lead of Solarize Washington, several similar "Solarize" efforts have sprung up around the state. In large part, these efforts are led by local installers and thus do not utilize the grassroots organizing central to a traditional Solarize program. Some of these campaigns have been held in towns not already hosting a Solarize campaign, while others advertise "rock

bottom prices" to compete with ongoing efforts in Solarize communities. Although these campaigns deviate from the established Solarize model, they do demonstrate the market-expanding effect that Solarize programs can have on the local solar industry.

For more information: Alexandra Sawyer, Northwest SEED, alex@nwseed.org , (206) 457-5403

Solarize Massachusetts: Spurring Competitive Installations

Campaigns: 4 (Harvard, Winchester, Hatfield, Scituate)
Installations: 162
Total Installed Capacity: 829 kW

Massachusetts Clean Energy Center (MassCEC) launched the Solarize Massachusetts (Solarize Mass) program in 2011 in an effort to expand the state's solar PV market beyond traditional early adopters. Working with the Massachusetts Department of Energy Resources, MassCEC identified four "Green Communities" to pilot the Solarize Mass program. To help spur grassroots marketing and activism around the program, MassCEC provided each community with an outreach toolkit – complete with a banner, yard signs, bumper stickers, templates, and other marketing materials – and framed the program as a competition to achieve widespread solar adoption. Though not the largest community in terms of population, the town of Harvard ultimately succeeded in installing the most systems and Solarizing an impressive 4% of total residences. Participants were able to choose between a direct ownership and a leasing ownership model, both with four tiers of pricing based on the number of people who contract to install solar. Although uptake of owned vs. leased solar installations varied by town, offering both options ultimately made the solar market more accessible for all.

Lessons and Considerations:

- *Each community is different:* MassCEC had a unique opportunity to learn from pilot projects that took place simultaneously in four distinct communities. Although each town selected for participation had been pre-designated as a "Green Community," each was unique in size and demographics. MassCEC provided the towns with the same outreach toolkit and found that the efficacy of the outreach tools varied significantly from town to town. While the Solarize model can be streamlined and standardized to a certain extent, it is ultimately a dynamic program that must be customized to fit the character of the community it is serving.

- *Steep pricing tiers are motivational:* Solarize Massachusetts featured a tiered pricing structure for each of its four pilot campaigns. Communities selected different installers, so pricing tiers differed between communities. In Harvard, the installer offered aggressive pricing tiers with sizable savings from tier to tier. This translated into momentum for community members to get their neighbors to participate—more participation meant substantial price savings for all. MassCEC learned that communities want early wins and demonstrable opportunities to save and showed aggressive price reductions based on group pricing are one way to accomplish this.

For more information: Elizabeth Kennedy, MassCEC, ekennedy@ masscec.com , (617) 315-9321

Vermont Solar Communities (Solar PV): Creating a For-Profit Spinoff

Campaigns: 10 (Waterbury 2, Williston 2, Charlotte 2, Hinesburg 2, Shelburne 2)

Installations: 60 solar PV

Total Installed Capacity: 300 kW

Vermont Public Interest Research Group (VPIRG) created VPIRG Energy to orchestrate solar group purchasing. Beginning in 2010, VPIRG Energy launched its Solar Communities program in three rounds—the first round offering only solar PV and subsequent rounds offering either solar PV or solar hot water. As an established membership-based organization, VPIRG had a pool of members from which to draw participants, so the Solar Communities effort required less organizing by local community groups than a traditional Solarize campaign might. Local community members with solar installations were particularly effective in organizing house parties to publicize the Solar Communities program and increase awareness about solar. VPIRG Energy recouped its operational and program costs through a $0.25 per watt lead-generation fee paid from the installer to VPIRG Energy at the time of system installation. The co-directors of VPIRG Energy concluded their campaigns and launched a new residential solar enterprise in 2012. SunCommon is a for-profit company that offers homeowners a solar lease for no money down.

Lessons and Considerations:

- *Sticking with one installer:* Although VPIRG Energy performed a competitive selection process for the first iteration of Solar Communities, it opted to use the same installer again for subsequent iterations of the program instead of issuing an RFP. This alienated some smaller local contractors, who felt that they were blocked from participating in the expanding solar markets in the four pilot communities. Concerned installers asserted themselves as solar designers, not simply "wrench turners," and wanted the opportunity to be included to the fullest extent possible.
- *Using a lead-generation fee:* To cover costs associated with program administration, lead generation, and marketing, VPIRG Energy collected a $0.25 per watt fee (approximately 5% of the installed price) from the contractor upon close of sale. After refining the sales process to ensure cash flow, the fee was high enough to help VPIRG recoup costs while low enough to preserve the program price advantage.[7]

Vermont Solar Communities (Solar Hot Water)

Campaigns: 2 (Montpelier, Addison County)
Installations: 175 to date

Following the first round of its Solar Communities program, VPIRG Energy expanded beyond PV to pilot two solar hot water campaigns. These programs followed many aspects of the Solarize model and used a local company for manufacture and installation of solar hot water systems. Solar Communities that offered solar hot water were ultimately more successful than those that offered solar PV in terms of final participation numbers. This could be attributed to several factors, including lower upfront cost, the opportunity to purchase locally manufactured systems, and an impending incentive drawback that spurred action.

For more information: Colleen Thomas, Colleen@vpirg.org, (802) 223-8421

One Block off the Grid

Campaigns: 20 communities in 11 states
Installations: 1,500
Total Installed Capacity: 8 MW

One Block Off the Grid (1BOG) is a San Francisco-based for-profit company that aims to figuratively take city blocks "off the grid" through solar installations en masse. 1BOG establishes programs in target cities with promising solar markets and aims to address three major barriers to widespread solar energy implementation: 1) high cost, 2) confusing process, and 3) lack of trust between potential customers and installers. 1BOG's city-based programs feature ongoing solar campaigns that run for three months at a time. The group recently launched several new Web tools with the intention of making solar simple and accessible. 1BOG's US Solar Markets Map (www.1bog.org/nation) showcases solar PV demand by county and assigns state solar policy "grades," while its solar estimating tool (www.1bog.org/solar-estimate) allows customers to use mapping technology to see what a solar system might mean for their home.

Lessons and Considerations:

- *Using a per-watt fee*: To finance their business model, 1BOG charges a $0.25 per watt installed fee to the chosen contractor. This adds approximately $1,250 to the price of a 5 kilowatt system, about 5% of the overall system cost. The customer never "sees" this fee, because it is built into the flat-rate price offered by the contractor.
- *Remote solar site assessment:* 1BOG recently developed a program to design solar systems remotely though Web-based mapping technology. Using this technology, the organization is able to offer similar services as one might receive in a traditional site evaluation, but cut down on time by performing the evaluation over the phone. Using this technology, 1BOG allows customers to sign contracts remotely, thus creating efficiencies in the solar installation timeline.

For More Information: 1BOG, www.1bog.org/

Reduce then Produce: Home Performance Before Solar

Campaigns: 1 (San Diego)
Installations: 7 (in addition to 11 home-performance upgrades)

The San Diego-based California Center for Sustainable Energy (CCSE) administers the California Solar Initiative rebates in the San Diego area and works with home performance contractors under the Energy Upgrade California program. In summer 2011, CCSE launched "Reduce then Produce," a pilot program to integrate the two efforts. Homeowners were required to get a Home Energy Rating System (HERS) rating of 100 or lower, or get a home assessment and conduct an upgrade before going solar. Rebates were available for both the HERS rating and the energy upgrades. GroupEnergy (see GroupEnergy: Workplace Campaigns for Employers and Employees) guided the RFP for solar PV, selecting two installers. CCSE invited a pool of 30 qualified home-performance contractors to join the program, and 13 ultimately signed on. Although participation in the pilot was light, program managers are eager to refine the program design and try again.

Lessons and Considerations:

- *Home Performance Contractors Want Solar Business*: It is relatively common, at least in strong solar markets like San Diego, to find home-performance contractors who offer holistic solutions including solar PV. Because the campaign used pre-selected solar contractors, the home-performance contractors did not embrace the program as strongly as they might have. Using a single contractor for both the home performance and the solar installation might increase contractor enthusiasm and sales.
- *Homeowners Want Recognition of Their Home-Performance Efforts*: By requiring homeowners to prove home performance (through a rating or upgrade implementation) before going solar, the program engendered resistance in homeowners who felt that they had already improved home performance. Program managers suggest that a more appealing approach is "Reduce AND Produce," encouraging homeowners to do both, but allowing homeowners to self-certify that they have improved home performance. Especially in the relatively mild climate of San Diego, home-performance improvements may not return enough savings to risk losing the solar opportunity.

For more information: Siobhan Foley, California Center for Sustainable Energy, Siobhan.foley@energycenter.org

GroupEnergy: Workplace Campaigns for Employers and Employees

Campaigns: Multiple corporate and government clients
Installations: NA
Total Installed Capacity: NA

Founded in July 2011, GroupEnergy delivers collaborative procurement programs to make residential and commercial clean energy solutions simple, social, and cost effective. GroupEnergy designs and administers procurement programs to pool the buying power of an organization's workforce, securing lower purchase pricing and attractive terms for residential solar and energy efficiency improvements. This new employee engagement tool streamlines the process of researching and choosing the right vendors to help participants lower their utility bills and generate renewable energy. Employers see increased employee engagement in their corporate social responsibility and sustainability initiatives, while helping their community achieve its greenhouse gas reduction goals.

Each GroupEnergy campaign includes a competitive RFP process to evaluate and select vendors. As of early 2012, multiple campaigns had completed vendor evaluation and launched enrollment. GroupEnergy can work either directly with a large employer, or deliver partnership programs through an umbrella nonprofit or government organization. Current GroupEnergy clients include Adobe's Green Team, Bank of America, the Bay Area Climate Collaborative SunShares program, Genentech, and ICLEI - Local Governments for Sustainability USA. Under ICLEI, GroupEnergy is administering "Energy Benefits," a new clean energy procurement program offered to member communities as a tool for achieving climate action goals. Energy Benefits offers residential solar and energy efficiency solutions, as well as commercial facility solar aggregation on behalf of employer organizations. With its nationwide reach, GroupEnergy is reducing soft costs, securing favorable pricing and leveraging the workplace to bring solar to new markets.

For More Information: Jessie Denver, GroupEnergy, info@mygroup energy.com

GENERAL LESSONS AND CONSIDERATIONS

The following lessons and considerations are based on the feedback from all of the Solarize campaigns in this guidebook.

Tap the Grassroots

Solarize campaigns are successful because they tap the grassroots to design and market the program. In a positive feedback loop, the process of creating and deploying the program builds community pride that encourages higher levels of participation in the community.

> *"Solarize Southwest was perhaps the single most satisfying project I've worked on at SW Neighborhoods. The shared experience of residents attending workshops together and installing solar energy equipment at the same time helped to create a strong sense of community amongst those who participated, and helped us fulfill our mission to the community: to empower citizen action to improve and maintain the livability of southwest neighborhoods."*
>
> Leonard Gard, Project Coordinator, Southwest Neighborhoods, Inc.

Involve the Community in Decision Making

The RFP process is an opportunity for the community to create an empowering statement of values. With guidance from technical experts, volunteers craft the contractor selection criteria and exercise choice in the selection of the installer(s).

Use Community-Based Marketing

Solarize is a classic example of community-based social marketing: Information reaches people through face-to-face encounters with friends and neighbors, house parties, and other social interactions. Although the campaign uses the Web as well as traditional media, the thrust of the marketing appeal is personal. In contrast to a plea from the government or the utility, the appeal comes directly from a friend or neighbor.

Collaborate with a Trusted Local Organization and Assign a Project Manager

A successful campaign collaborates with a trusted local organization that has a history of helping people. In Portland, the Neighborhood Coalition offices served this role. In Washington, it was Northwest SEED, and in Vermont, it was VPIRG that was the trusted organization. Local organizations provide "third-party validation," which instills trust in the program. Regardless of the organization, each campaign had a dedicated project manager to orchestrate the effort.

Plan for Success

The first Solarize Portland effort set a goal of 25 installations. When 350 residents signed up, the manual process of entering enrollee information into a spreadsheet quickly became untenable, and the contractor realized that it needed a customer service plan to keep in touch with customers over the several months that they would have to wait to get through the installation queue.

Project organizers should plan for success and put efficient systems in place for capturing enrollment information, sharing information with contractors, and following up with customers. Consider selecting more than one contractor, so that no single contractor is overwhelmed with jobs.

Support Contractor Systems

Smaller contractors in particular may need support to build their customer relationship management systems to handle a program of regular follow-ups to keep warm leads "warm" until they can reach the customer. Project organizers can help contractors by ensuring that they have thought through their customer service plan, requiring specific plans in the RFP response.

Make Contractors Responsible for Site Assessments

The early Solarize campaigns offered an optional free site assessment in which Energy Trust helped residents determine their homes' suitability for solar and consider energy efficiency options. Although attractive in principle, in practice, offering these third-party reviews created a bottleneck, slowing the installation process as contractors had to wait for the reviewers to complete their assessment before meeting with the homeowner. The first program evaluation showed that homeowners who requested the optional site assessment actually installed solar at a lower rate than those who did not (possibly because they were requested by homeowners who suspected that their home was unsuitable.) In any case, the contractor must ultimately visit the home to advise on the system size and sign the contract, so the site assessments can be part of the contractor's plan.

Pricing Considerations

To what extent is the success of Solarize due to low prices? Campaign results suggest that prices and incentives vary widely from market to market, and project organizers should consider several points when designing the price of the offer.

Absolute Price Is Less Important Than the Perception of a Good Deal

In general, most people don't know what a solar installation is supposed to cost, so they have no price yardstick to evaluate the program offering. More important than getting a good deal is the assurance that they are not getting a bad deal. As long as a consistent price is set for everyone, and it is demonstrably less expensive than the "going rate" for individual solar installations, people perceive the cost as "a good deal." In fact, many RFP committees selected final bids that were not the lowest price, but the best value, providing a reasonable price for high-quality service.

Fixed Price vs. Descending Price: Pros and Cons

The Solarize campaigns in Portland and Massachusetts effectively used a descending price scale to encourage higher participation. While a descending price can motivate early enrollees to recruit others, it also adds complexity: it delays the time when the final price is determined, so the contractor cannot quote a final price to early enrollees. Contractors might quote the highest price and collect payment in several installments, with a contract clause that the final installment will be adjusted to reflect the final price. However, organizers might consider fixed-flat pricing from the start and use other means to encourage recruiting. For example, contractors might pledge to donate a system to a local organization if the installations reach a certain goal. Others may offer a rebate at the end of the program to all participants based on the total solar capacity installed.

Another argument for fixed pricing is that the contractor's ability to offer a lower price does not depend as much on the savings on volume purchases of equipment as it does on the savings in time and effort in marketing. Larger contractors often have access to volume equipment pricing even without the group purchase, so their savings are more likely to be realized in the community-run sales and marketing. They can commit to their lowest price knowing that the grassroots community-based social marketing effort will bring them hot leads with a high conversion rate.

Financing

By offering some form of program financing, Solarize campaigns are able to tap a larger market for PV. Options vary from state to state, but some combination of the following should be considered.

Municipal Loans

The City of Pendleton, Oregon, offered zero-interest loans of $9,000 to Solarize participants. Funds were borrowed from an existing wastewater treatment facility rate stabilization fund, with loan repayment structured over four years: 50% paid back the first year and the remainder paid back over three years. The loans were secured by a lien against the property through a Local Improvement District. The City used property tax revenue to make up the lost interest on the rate stabilization fund, so that when the loans are fully repaid, the fund will be exactly where it would have been without the program. These funds bridged the gap between the customers' payments to the contractor and their receipt of state and federal tax incentives. Ultimately, 63% of residential participants took advantage of the loan. Sixty-four homeowners and several small businesses used $775,000 in loans. One year later, none of the loans are in default, and the City has recovered 50% of what it loaned out and is on track for 100% recovery of the loan fund. The loans have given such a boost to solar in the area that Solarize Pendleton's selected installer, Livelight Energy, has opened a permanent office in the rural area.

For more information: City of Pendleton, http://solarizependleton.com/main/replication

Bank or Credit Union Loans

As the residential solar market continues to expand, lenders are beginning to realize the value of tapping this market. For example, Portland-based Umpqua Bank offers "Greenstreet" lending, a menu including a home equity line of credit, home equity loan, or unsecured loan to homeowners who are improving energy efficiency or going solar. The rates are competitive, and the loans have the added advantage of no fees and a lender that understands the value of connecting with green neighborhood initiatives. Credit Unions may also offer special loans for members or employees who are going solar, as the San Jose Credit Union did for its employees in 2010.

Solar Leases or PPAs

In certain markets (depending on eligibility for state and utility incentives and the ability to legally offer third-party financing solutions), companies may offer customers the option to lease panels rather than make a direct purchase. In Portland, when state tax rules changed to allow homeowners to take the state tax credit regardless of ownership, two large solar companies, SunRun and SolarCity, introduced financing alternatives to direct ownership. SunRun partners offer a "Prepaid PPA" in which the customer pays $6,000 up front for a 3.24-kW system and recoups the payment over the next four years as a state tax credit. Customers receive 20 years of electricity from the panels on their roof, with maintenance and warranties covered by their installer. Solar City offers a 15-year solar lease, with no upfront payment, for as little as $20 a month.

Third-party prepaid and lease solar PV options have provided more choice for homeowners, with lower costs. Nine out of 10 homeowners choose these programs over standard purchase options.

Utility Loans

Until recently, there have been few utilities offering loans for going solar, and no overlap between solar loans and Solarize campaigns. Now, in Washington, Snohomish County Public Utility District No. 1 offers its Solar Express loan and rebate program in conjunction with a Solarize campaign. Under Washington law, Snohomish can claim double credit toward meeting the requirements of the state Renewable Portfolio Standard because the kilowatts installed are distributed renewable generation.

Program Funding Considerations

Deploying a Solarize campaign costs money. Despite harnessing volunteer labor for everything from planning to marketing to contractor selection, a successful campaign will need the oversight of a project manager and will incur costs for marketing materials, database administration, and communications. The Portland campaigns relied on the staff at the neighborhood coalitions, as well as paid staff from Energy Trust and City of Portland, who were supported in part by a federal grant. Communities without paid neighborhood staff or grants should consider other options for funding.

Collect a Per-Watt Fee

Project organizers could consider building a small per-watt fee into the contractor's scope of work, which is then passed on to the customer. The contractor can still offer a competitive price, because it is saving money on marketing, while the program establishes a source of funding for everything from staff time to outreach materials. Organizers might still need seed funding to launch the project until the installations and fees begin to flow.

Create a Buyer's Co-Op

Salem Creative Network adopted a co-op model to fund its Solarize efforts, charging program participants a fee of $0.10 per watt (e.g., $250 for a 2.5-kW system) to join the co-op. The fee was intended to cover program management, database administration, and outreach. The co-op fees supported the campaign organizing staff for about a month. It may be more palatable to customers to have the co-op fee rolled into the contractor fee, so that they only write one check.

Leverage Contractor Marketing Dollars and Expertise

As noted, the community-led marketing campaign saves contractors money. In return, the selected contractors may have marketing materials and expertise that they can share with the campaign. For example, in Pendleton, Oregon, the contractor provided yard signs and marketing fliers, rented a booth at the farmer's market, and covered other incidental marketing costs. In Seattle, installers have provided financial assistance with everything from door hangers to bus and radio advertisements.

Secure Local Utility Program Funds

Electric utilities may be interested in supporting the labor costs and/or the rebates for a Solarize campaign as a way of delivering a popular customer service. The campaigns provide outreach and education about energy and build a constituency that interacts more closely with the utility. In addition, utilities may be able to claim Renewable Energy Credits (RECs) from programs that they sponsor. For example, in Washington, utilities that incentivize solar PV can double count that production toward meeting their Renewable Portfolio Standard.

PLANNING A SOLARIZE CAMPAIGN

The following section describes the steps to carry out a successful Solarize campaign.

Step 1a: Develop Partnerships and Initiate Planning (Months 1 – 3)

A successful campaign begins with strong planning and partnerships. The institutional project organizer should enlist key allies and support starting with a primary project manager (one very dedicated individual who will oversee all the moving parts). Usually, the initial campaign organizing involves these players:

- Primary project manager (institutional project lead)
- Community organizer or volunteer coordinator (may be a neighborhood volunteer)
- Technical support lead (a solar specialist, such as utility or city staff)

These project players collaborate to build the project work plan and timeline, identifying all the tasks, responsible parties, and community partners. Potential community allies include:

- American Solar Energy Society chapter
- Local nonprofit
- City government

- Local utility
- Neighborhood coalitions or associations
- Local manufacturer of solar equipment
- Churches
- Rotary or other service clubs
- Credit union or local bank

Sample Project Timeline

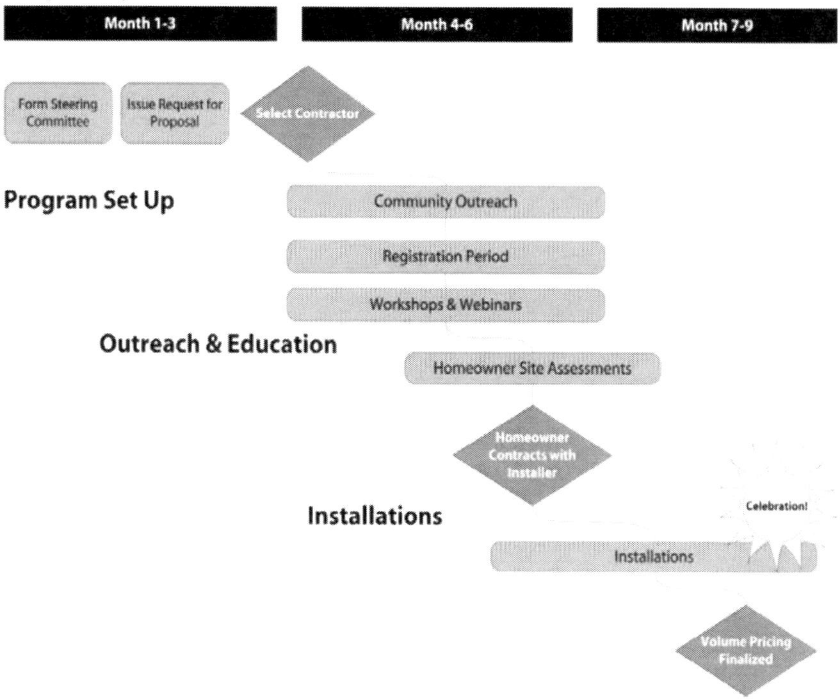

Sample Roles in a Solarize Campaign

The following chart shows sample roles and responsibilities in a typical Solarize campaign. The Project organizer is an essential role and could be a neighborhood coalition, a municipality, a local ASES chapter, or any organization with the capacity to devote a half-time person to leading the charge.

	Project Manager	Volunteers	Contractor	Utility/Municipality
Planning	Manage program; provide resources	Provide ideas		Provide tech support; provide resources
Volunteer Recruitment	Recruit & organize committees			Advise committees
Request for Proposal (RFP)	Issue RFP; advise on RFP and contractor selection	Draft RFP; select contractor	Respond to RFP	Advise on RFP
Outreach	Manage outreach campaign; create and print fliers; lead workshops	Build website; distribute fliers, outreach materials; schedule workshops; identify venues	Teach nuts and bolts and Q&A session	Provide workshop curriculum; teach workshops
Enrollment	Compile database of enrollees; engage customers	Recruit neighbors	Conduct preliminary assessment and schedule site assessments	
Site Assessments	Track contractor turnaround time and signed contracts		Conduct site assessments with homeowners; prepare bids	
Installations	Track contractor turnaround time and customer experience		Execute contracts; install systems; complete paperwork	Streamline solar permitting process; inspect installations; interconnect systems
Celebration!	Issue press release; promote, evaluate and replicate	Plan and/or host party	Plan and/or host party	Evaluate

Step 1b: Build Database and Customer Interface (Months 1 – 3)

A customer service database is essential for contractors and project organizers to track customer follow-up, schedule installations, and capture project results. The project organizer should provide the database structure and protocols to the contractor. Some contractors may have their own customer relationship management (CRM) software, but they should also update the database supplied by the program. This way, the contractor's process is transparent to the project organizer, and if there are delays in implementation, the project organizer can see these and plan accordingly. Solar Oregon has developed a database for use with Solarize projects and is available on contract to build, manage, and administer a customized database for a reasonable fee. Other Solarize projects have used CRM software such as Salesforce to manage customer enrollment. The project organizer, as the agent of public trust, must be sensitive to the privacy of participants, and take care not to disclose information beyond the program or misuse information submitted by participants.

Step 2: Volunteer Recruitment (Months 1 – 2)

One of the first tasks of the primary project manager is to host a meeting to recruit core volunteers. The core volunteers can be organized into two committees:

- Outreach committee: Manage the communication and outreach to all neighbors. Members should be media savvy and should get articles in the newspaper, build a website, and recruit neighbors.
- RFP committee: Write the RFP, review contractor submittals, interview, and select a contractor. Members should include at least one solar professional or tradesperson and preferably non-voting technical support from the institutional sponsor (e.g., city or neighborhood coalition).

The neighborhood recruitment meeting should be advertised in neighborhood papers, discussed personally with neighborhood association chairs/ community groups, and talked about widely. These volunteers will be the core group to initiate the Solarize effort.

Step 3: Request for Proposal Process (Months 2 – 3)

Writing the RFP and creating the RFP scoring rubric is a chance for community members to express their values. The volunteer committee, acting as the agent of public trust, is creating a defensible, open process to select the contractor. Usually, at least one solar professional or person with solar expertise supports the committee. It is important to have a clear method of scoring the proposals and to communicate this to the bidders. It is best to use a low number of points (three to five) for each desired category, so that score variations are significant.

"The RFP process was extremely important for our committee. We learned more about the contractors than we ever could as individual customers, and we communicated our values to the contractors."

Todd Farris, Volunteer Program Manager, Solarize Southwest

The RFP should be issued widely. Contractors should have several weeks to respond, and all questions and answers should be posted to a public website, where all bidders see the same information. After proposals are received, the committee members begin evaluation. Even where the project receives many excellent proposals, it is best to interview only the top two or three contractors as a courtesy to the volunteers and contractors.

After contractor selection is announced, the committee may receive inquiries from contractors who were not selected. If they have followed the RFP evaluation criteria, the response is simple: "The committee scored the applicants and chose the one or two that scored the highest on the rubric." Committee members should keep conversations positive and not try to justify why the committee chose one contractor over another.

Step 4: Outreach and Education (Months 4 – 6)

Once the contractor is selected, outreach becomes a focus. The outreach committee creates or adapts materials – fliers, buttons, stickers, yard signs, and a website to help spread the word. Elements of the outreach campaign can include:

Website

A program website serves as a central location for updates on the campaign, a calendar of events, and a place to enroll. It is essential for volunteers to direct people to the website for timely information. Having a dedicated volunteer to update the website regularly helps build and maintain program momentum.

Print Materials

A colorful campaign logo and photo on a flier help lend legitimacy and spread the word. Fliers, posters, door hangers, and other print materials should be distributed widely.

Blogs and Emails

Electronic media provide an affordable and convenient way to increase the outreach of the campaign. The outreach committee should submit information to neighborhood blogs, write letters in their neighborhood newsletters, and send emails to friends, neighbors, and family members encouraging them to join the campaign.

Workshops

All interested homeowners are strongly encouraged to attend at least one workshop. A contractor representative should attend each workshop to answer questions. This will provide technical support to workshop presenters, while building a relationship of trust between the contractor and the homeowners. The group setting is important, to build trust and neighborhood cohesion while encouraging attendees to enroll in the program.

Basic Workshop

This is an introductory, one-hour workshop, held at multiple locations throughout the community. The basic workshop explains how the project works, the benefits of collective purchasing, solar PV technology, financing and incentives, and how to participate.

Technical Q&A Sessions

If participants want additional, in-depth information, organizers may consider holding technical Q&A sessions. These informal, open-format sessions allow potential participants to get their questions answered in a friendly and educational environment. Sessions could focus on a topic presented by subject matter experts:

- Cash incentives, tax credits and financing (Presenter: financing partner/utility)
- Net metering (Presenter: utility)
- Technical nuts and bolts (Presenter: contractor)

Solar Ambassadors

A successful campaign will enlist the support of solar champions who already have solar on their homes. For example, Solar Oregon organizes a program of Solar Ambassadors, local residents who have gone solar. These supporters are strong advocates and positive examples for homeowners considering a solar purchase. Ambassadors can attend or present at workshops, providing an important validation to others looking to install solar.

Step 5: Customer Enrollment (Months 4 – 6)

The enrollment period, usually three months, should run concurrently with outreach and education. Kick off with a press release and a high-profile community event, perhaps at a farmer's market or other public venue. Ideally, enrollment occurs online, and participants enter their data directly into a database. Programs may make a provision for participants to register by phone if they have no Internet access, and a project organizer could enter this customer data into the Web interface. The online enrollment process should generate an auto-reply email, alerting the customer of the date on which their information will be given to the installation contractor, and telling them to expect a call within two weeks (or the agreed upon turnaround time). At this point, the leads are hot and the sooner the contractor can act, the more likely the leads will convert to installations.

Throughout the enrollment period, the outreach committee volunteers drive people to the website through various avenues that suit their own comfort level: hosting coffees, going door to door, sending emails, posting fliers on public message boards, or submitting articles to the local press. As the enrollment period draws to a close, the media may take interest, if they haven't already. It is best to invite media early on, so that they can help get the word out, rather than generate a lot of interest after the enrollment has closed.

Step 6: Site Assessments (Months 4 – 8)

As soon as people begin enrolling, the project organizer can begin passing participant information to the contractor. Although several Solarize campaigns waited until the end of enrollment to pass the leads to the contractor, passing leads as participants enroll will help even out the contractor workload and improve the follow-up time. The contractor may use mapping software to screen out any obviously ineligible participants (e.g., those with heavy shading) and then schedule an appointment to meet with the customer for a more detailed evaluation and system sizing. If all goes well, the customer and contractor sign a contract for installation.

Step 7: Installations (Months 5 – 9)

The contractor is responsible for installations, but the project manager should monitor the customer database to ensure that installations are occurring within an appropriate time frame. At this phase, the contractor should be updating the customer database as customers are contacted and systems installed. All customers should continue to get periodic messages from the program, offering updates on the status of the program. In Portland, the project manager coordinated weekly or twice-monthly team meetings to discuss installation statistics, and address and issues or concerns that arose. Meetings built a strong team atmosphere and gave the City, neighborhood leads, and the contractor opportunities for increased project cooperation and correction, when needed.

Some Portland participants expressed frustration with long waiting periods between enrollment and installation. This is typical in a volume purchasing program, but can be alleviated in part by choosing more than one contractor and/or releasing names to the contractor as soon as homeowners enroll.

Step 8: Celebrate and Reflect (Month 9)

It is important to acknowledge the hard work of everyone who supported the program and celebrate the community effort. The contractor and/or manufacturer may be willing to sponsor a public celebration. The media will want to attend, and the positive energy generated by the celebration can help fuel the next project in the next neighborhood. Equally important is reflection and evaluation. Project organizers can continue to build public trust by listening to feedback in order to improve future programs.

> *After the installations were complete, the homeowners came together for a walking tour of neighborhood homes and a celebratory picnic. Another Portland neighborhood held its celebration at the local brewpub.*

Sample Budget

Although every program will vary by location and population size, this sample budget provides a starting point for project planners.

The budget above reflects a possible scenario for a project lead, volunteers, and program partners. Of course, labor costs will vary widely, depending on how much of the labor is volunteer based.

Some communities have successfully leveraged AmeriCorps or other service-learning volunteers to serve key program roles, while others have used volunteers primarily in outreach and the RFP process. In some municipalities, the existing staff in the office of neighborhoods or the office of energy or sustainability can take on the project lead hours as part of a special campaign.

As with labor, the materials budget will vary widely, depending on the media market and the amount of outreach materials that can be donated. The budget does not suggest a source for program funding. Each program planner will have to consider options discussed elsewhere in this guide, including grants, volunteer contributions, or a fee assessed on each installation.

Labor Hours	**Project Organizer**	**Volunteers**	**Contractor**	**Utility**	**Total**
Project Management	250	250			
RFP Committee	40	80			
Outreach Committee	70	50			
Workshop Design and Delivery	100	50	20	10	
Site Assessments			*		
Installations			*		
Celebration and Evaluation	30	30	20		
Total Hours	***490***	***460***	***40***	***10***	***1,000***
Materials Expenses					
Collateral (fliers, yard signs, etc.)	$1,000		$1,000		
Advertising	$150				
Database Development	$2,000				
Labor Hours	Project Organizer	Volunteers	Contractor	Utility	Total
Workshop Venue Rental	$400				
Speaker Fees	$300				
Booth Rental for Events	$100				
Web Hosting/Domain Name	$200				
Celebration Event	$200		$300		
Total Materials	***$4,350***	***$0***	***$1,300***	***$0***	***$5,650***

* Contractor hours for site assessments and installations will vary by number of participants and are not shown here because they are not unique to a Solarize campaign.

Additional Resources and References

Collective Purchasing Programs from Across the United States

[1] GroupEnergy launched multiple collective purchasing programs for the workplace in early 2012. www.mygroupenergy.com

[2] "Lighten Our Load" was developed for Columbia Sportswear by Energy Trust of Oregon in 2008. www.energytrust.org

[3] MadiSUN Group Solar Program serves residential and commercial customers in Madison, Wisconsin. http://www.cityofmadison.com/sustainability/city/madisun/

[4] Make Mine Solar is a collective hot water purchasing program, based in Minneapolis, Minnesota. www.mnrenewables.org/MakeMineSolar

[5] One Block Off the Grid is active in 20 cities nationwide, supporting volume purchasing for residential customers. www.1bog.org/

[6] Reduce then Produce was an integrated home performance and solar program based in San Diego, from the California Center for Sustainable Energy.

[7] San Jose Employee Solar Group Buy was offered to City employees and retirees in 2010. The program became the model for the SunShares Program of the Bay Area Climate Collaborative. http://baclimate.org/impact/sunshares. html

[8] Solar Beaverton offers PV, hot water, and even EV charging stations in a group purchase format. http://www.beavertonoregon.gov/index.aspx?NID=412

[9] Solarize Eugene: The Resource Innovation Group won a utility grant for Solarize Eugene 2012, addressing PV and hot water. http:// solarenergy design.com/solar-electric-systems/solarize-eugene/

[10] Solarize Pendleton: The City of Pendleton, Oregon, offered zero-interest loans to finance solar installations and created program replication materials http://solarizependleton.com/main/replication/

[11] Solarize Portland: With over six campaigns and 560 installations, Portland leads the way and helps other cities run Solarize campaigns. http://www. portlandonline.com/bps/index.cfm?c=51902

[12] Solarize Salem: The Salem Creative Network organized a co-op to help fund its PV and hot water campaigns. http://solarizesalem.org/

[13] Solarize Santa Barbara: More than 49 neighbors went solar through a program from Community Environmental Council. http://www.cecsb.org/solarize-santa-barbara

[14] Solarize Washington: A series of residential PV campaigns from Northwest SEED began in 2011. www.solarizewa.org

[15] Solarize Massachusetts: Massachusetts Clean Energy Center (MassCEC) in partnership with Green Communities Division of the Massachusetts Dept of Energy Resources ran campaigns in four cities. http://www.masscec.com/index.cfm/cdid/12093/pid/11159

[16] VPIRG Energy ran successful "Solar Communities" programs for PV and hot water across Vermont. The co-directors of VPIRG Energy have subsequently launched SunCommon. http://suncommon.com/

Publications

[1] **Solar Powering Your Community: A Guide for Local Governments.** U.S. Department of Energy, 2011. The second edition of this guide includes case studies and lessons learned from Solar America Cities. www.solaramericacommunities.energy.gov/resources/guide_for_local_governments

[2] **Solarize Portland: Community Empowerment through Collective Purchasing.** Lizzie Rubado, Energy Trust of Oregon, August 2010. This paper provides more details on the success of Solarize Portland. www.energytrust. org/About/policy

[3] **Evaluation of Energy Trust of Oregon's Solar Programs: Solarize Southeast Portland and Solar Energy Review.** The Cadmus Group, November 2010. The evaluation contains detailed customer feedback and participation profiles for the first Solarize project.www.solarizewa.org/files/cadmus-report-eto-solar-programs

[4] **Smart Solar Marketing Strategies.** L. Rosoff and M. Sinclair, Clean Energy Group 2009. The report offers valuable lessons in marketing solar.www.statesadvancingsolar.org/resource-center/smart-solar-marketing -strategies

Peer Effects in the Diffusion of Solar Photovoltaic Panels. Bryan Bollinger, NYU Stern School of Business and Kenneth Gillingham, Yale School of Forestry & Environmental Studies, December 2011. This paper analyzes the spread of solar PV in communities. http:// people.stern.nyu.edu/bbolling/index_files/BollingerGillingham_PeerEffects Solar.pdf

End Notes

[1] Tracking the Sun IV: The Installed Cost of Photovoltaics in the U.S. from 1998 – 2010, Lawrence Berkeley National Laboratory, Jan. 2011, p. 1. http://eetd.lbl.gov/ea/ems/reports/lbnl-5047e.pdf.

[2] Eighteen full-time permanent jobs were created by three solar installation firms (2010 self-reported numbers).

[3] The SAC grant has supported a half-time employee for two years. Additional City staff provided technical and management assistance. Portland has also provided sub-recipient grant funding to eight communities in Oregon totaling $47,000 with the intent of replicating the successes of Solarize Portland across the state.

[4] Estimate based on conversations with contractors in the Portland area in 2010.

[5] Avg. U.S. residential price: $6.24 from U.S. Solar Market Insight Reports, Q32011 Full Report, SEIA/GTM Research, p. 28.

[6] Although Portland defined "local" as "Made in Oregon," other campaigns may define "local" as "made in the region" or "made in the USA." Regardless of the definition, the desire to "buy local" can influence customer choices and boost the "local" economy.

[7] Presumably, as costs for solar PV continue to fall, customer acquisition costs and lead generation fees will fall.

INDEX

A

B

C

D

E

F

G

H

I

J

Q

R

S

Y